# TECTONIC RESONANT EVOLUTION

## A Bold New Theory on the Origins of Humanity

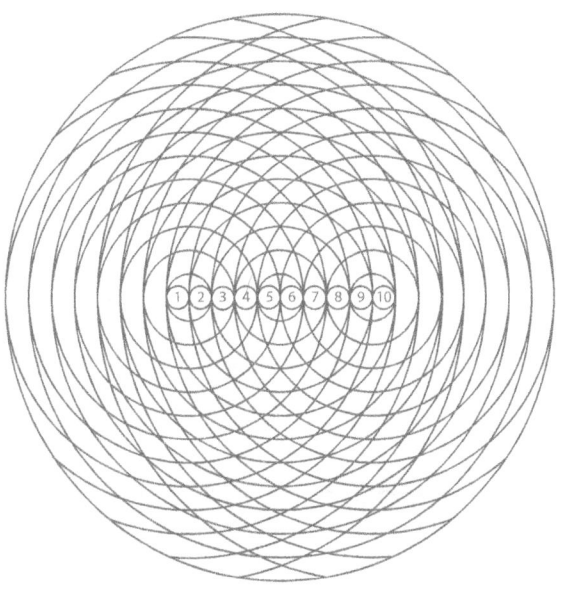

by
**Ben Aster**

BLUE HOUSE BOOKS
Baltimore 2020

Copyright © 2020 by
Blue House Books
All rights reserved.

Permission to reproduce any part of this book,
in any manner whatsoever,
must be obtained from the author.

Please direct all correspondence to
Blue House Books
P.O. Box 4852
Baltimore, MD 21211

Library of Congress Control Number 2019918245
ISBN 978-1-7344088-0-5

*Printed in the United States*

# **DEDICATION**

Future generations have the possibility to seek out truth, do more investigation, and review the evidence and ideas of current and previous generations. All the myths, sacred scriptures, folklore, teaching stories and libraries from ancient to current times deserve a finer examination from a fresh new perspective. We are fortunate, for in the past few centuries, much of the work has been done by previous researchers, historians, mythologists, linguists, anthropologists, archeologists, cartographers, and other truth seekers.

According to the authors of the Bible, Jesus says, "You are flesh of my flesh and blood of my blood." We are the flesh and blood of our parents and their parents before them. This book, we dedicate to everyone, for future generations.

The knowledge and wisdom of our past, of our ancestors, provides few clues for modern peoples to derive an understanding of their origins. Ancient texts, like the *Egyptian Book of the Dead*, the *Tibetan Book of the Dead*, and scriptures, affect the living, and the authors are all dead.

We give thanks for the seekers and revealers of truth, who struggled through many harsh conditions while entering into record any evidence relevant to the general quest for truth. As a result of their efforts we can share an entire new theory.

# CONTENTS

Prologue ..................................................... vii
Preface ........................................................ xi
Chapter 1: Resonance ................................. 1
Chapter 2: Tectonic Resonant Evolution ............... 7
Chapter 3: Flood Adaptations ........................ 9
Chapter 4: Ice Age Adaptations ..................... 11
Chapter 5: Evolution of the Human Skull ........ 13
Chapter 6: Template for the Human Skull ....... 17
Chapter 7: Lobotomy ................................... 21
Chapter 8: Poseidon ..................................... 25
Chapter 9: The Sacred Tortoise .................... 29
Chapter 10: Temple Builders ........................ 37
Chapter 11: Geodesy ................................... 43
Chapter 12: The Living Earth ....................... 47
Chapter 13: Being a Resonant Being ............. 51
Chapter 14: Resurrection ............................. 55
Chapter 15: Conclusions .............................. 59
An Autobiographical Afterword .................... 63

# PROLOGUE

Truth is said to be self-evident, self revealing, and law conformable. The facts speak for themselves. There are certain special pieces of information, defining moments of time, in which there is a major shift or change in the thinking and feelings of the people, that supplicates, gives, or provides them a new paradigm of self perception, individually and wholly. The most important information for people to discover fills them with a sense of self, a sense of identity forming the corpus roots of their family, tribe and ethnic sense of self.

Every age has its own unique characteristics, like a name for its peoples. The prophets saw what they believed to be the last age. From the ashes of the old arises a new age, a more aware kind of people.

The spiritual religious community believes we are specially created by God and do not believe in evolution. Evolution is taught in many different systems of education, which implies we are not created by God but by over a billion years of adaptation.

During my collegiate education I thought that evolution meant that with each new generation we are evolving, meaning we are better than the previous generations. Learning and building upon the skills of the previous generations, humanity becomes smarter and more capable. One evening I decided to tell my parents I believed I was more evolved than they were. Neither of them agreed. Shaking their heads from side to side, my parents responded, "So, here's your bill for college! You can pay it!" As evolved as I was, I was not smart enough to know better then to say to my parents, I am more evolved than you, in theory.

Artifacts of lost civilizations are found everywhere upon the surface of the Earth, even though supposedly we all originate from Africa. Oddities preserved, and mummification performed by the ancients, held their hope that future generations could know about their ancient relatives, in the story of the human species.

# PREFACE

Children have a natural curiosity, questioning everything around them, much like Socrates, whose maxim was "Know Thyself." Where do we come from? How did we get here? Who am I? There are many mysterious answers to our inner child questions in the form of origin stories, mythology, religion, and more. The stories told to children provide them a dreamscape of inner life and gives them the causes needed to seek out more stories discovering more truth. Our great questions should have great answers.

We introduce a bold new theory for the masses to chew on, the theory of Tectonic Resonant Evolution. It embodies an interdisciplinary approach towards answering the questions pertaining to the origins of humanity. From a new perspective, we contribute new evidence into the plethora of already existing ideas, concepts and theories, as to the origins of humanity. It will be for future generations to question or resolve these in their own time.

The theory of Tectonic Resonant Evolution blends the currently evolving anthropological arguments with the currently evolving special creation ideas. Usually these two ideas do not blend well at all in the mainstream schools of thought or the traditional religious institutional systems of belief. In this book we provide some of the first axioms, postulating that humanity's image is derived from God, which may come as a shock to the mainstream spiritual faiths, hardcore atheists, and hard working scientists. Did God have everything planned from the beginning, from before the beginning? If so, why was He so shocked when Adam and Eve suddenly realized that they were naked?

We could call this theory the mega-macro-geo-tectonic-resonant-terra-formal-auto-proto-genesis theory of evolution because it implies external structural forces are at work in crafting the human image, by terra-forming both of the major land masses largely existing south of the Equator.

All empires stylize their own version of the Celestial Palace where the ruling person sits on a royal throne symbolically representing the will of the Most High. Jesus Christ installs the blueprint for the creation of an Earthly Kingdom of Heaven, which humanity inherits from earliest tribal societies, between the tribal mother-queen and the tribal father-king. There is an essence of divine natural law, a living geometry of resonance in the patterns forming esoteric groups, civilizations, tribal societies and kingdoms.

# CHAPTER 1:
# RESONANCE

Oral story telling is an ancient tradition practiced by most cultures around the world. The keepers of the stories gather closely together in a circle. One member begins with a story the next member continues with the next story, around the circle, until the sacred wisdom of their ancestors has been recited. Ancient wisdom and knowledge has been passed down through the generations around the sacred circle. Mandala in Sanskrit means circle, defining the limits of unity and creation. The circle is as significant as the invention of the wheel.

Ancient scholarly orders often transfer their knowledge in secret messages and hidden ways. For example, an ancient scholar might pour sand onto a drum, pan, gong, or shield, whatever is available, then beat the drum either near the edge or in the middle. The rhythmic beats of the drum displace the sand in the form of a monopole. A monopole is one center of displacement, force, or cause of motion, forming a ripple effect. If a second drummer joins in, the sand displaces, forming a duality, a dipole, two centers of displacement.

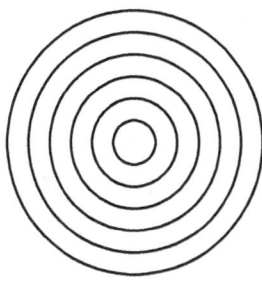

**Figure 1. Monopole**

*The vibrations, of one drummer, produced by the drumbeats appearing on the surface of the drum.*

There are two forms of duality, unblended and blended. Unblended duality is when the vibration of two drumbeats never meet, except for at the center where the two vibrations can touch. Blended duality is when the two vibrations overlap, cross sectioning each other. Many variations of unblended and blended duality exist in nature.

**Figure 2a.
Dipole Unblended Duality**

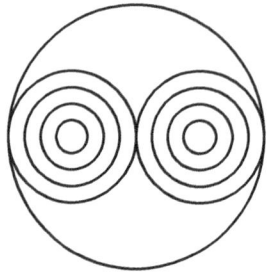

**Figure 2b.
Dipole Blended Duality**

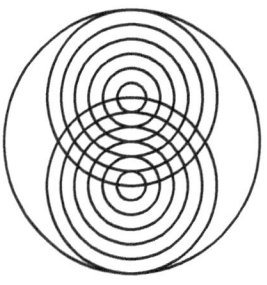

*The vibrations of two drummers, produced by the drumbeats, appear on the surface of the drum.*

Cymatics comes from the ancient Greek word kuma, meaning wave, and is the modern study of wave phenomena and vibration. The Chinese sprouting bowl is a bowl filled with water. When the edge of the bowl is rubbed it produces a humming sound, much like a crystal glass filled with water. The sound from the edge of the bowl is echoed back and forth along the circular edge of the bowl producing concentric rings in the form of ripples in the water. Resonance comes from the Latin root resonare, meaning to resound.

**Figure 3a. Action**  **Figure 3b. Reaction**

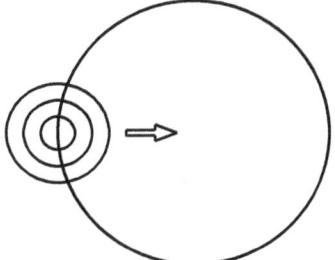  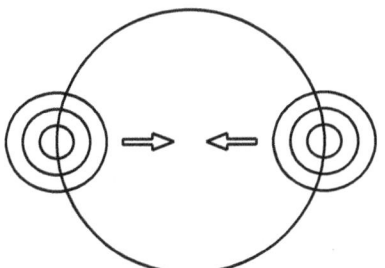

*Sound travels across the bowl from the edge of the Chinese Sprouting Bowl apparatus in the form of a Monopole.*

From one edge of the bowl to the other, sound echoes back and forth forming the Sacred Duality creating a third force, the Holy Trinity. The concentric rings emanating from the center are an example of the third force.

**Figure 4. The Concentric Rings of the Chinese Sprouting Bowl.**

Creation vibrates like the Chinese sprouting bowl. All matter vibrates at a consistent rate of frequency displacing all other matter. The resonant field of matter is the displacement of matter, and matter's consistent vibration is the resonant frequency, or natural frequency of matter.

In the Chinese sprouting bowl example, the bowl is a resonating field, the cause of the sound. The concentric rings rippling across the surface of the water proves resonance creates and maintains structure. Humming, the bowl produces a sound similar to the resonating field of creation, known in ancient cultures as the word of God, "Om," creator of all things, a rainbow of music, the "Song of Songs," love.

## Figure 5. Song of Songs

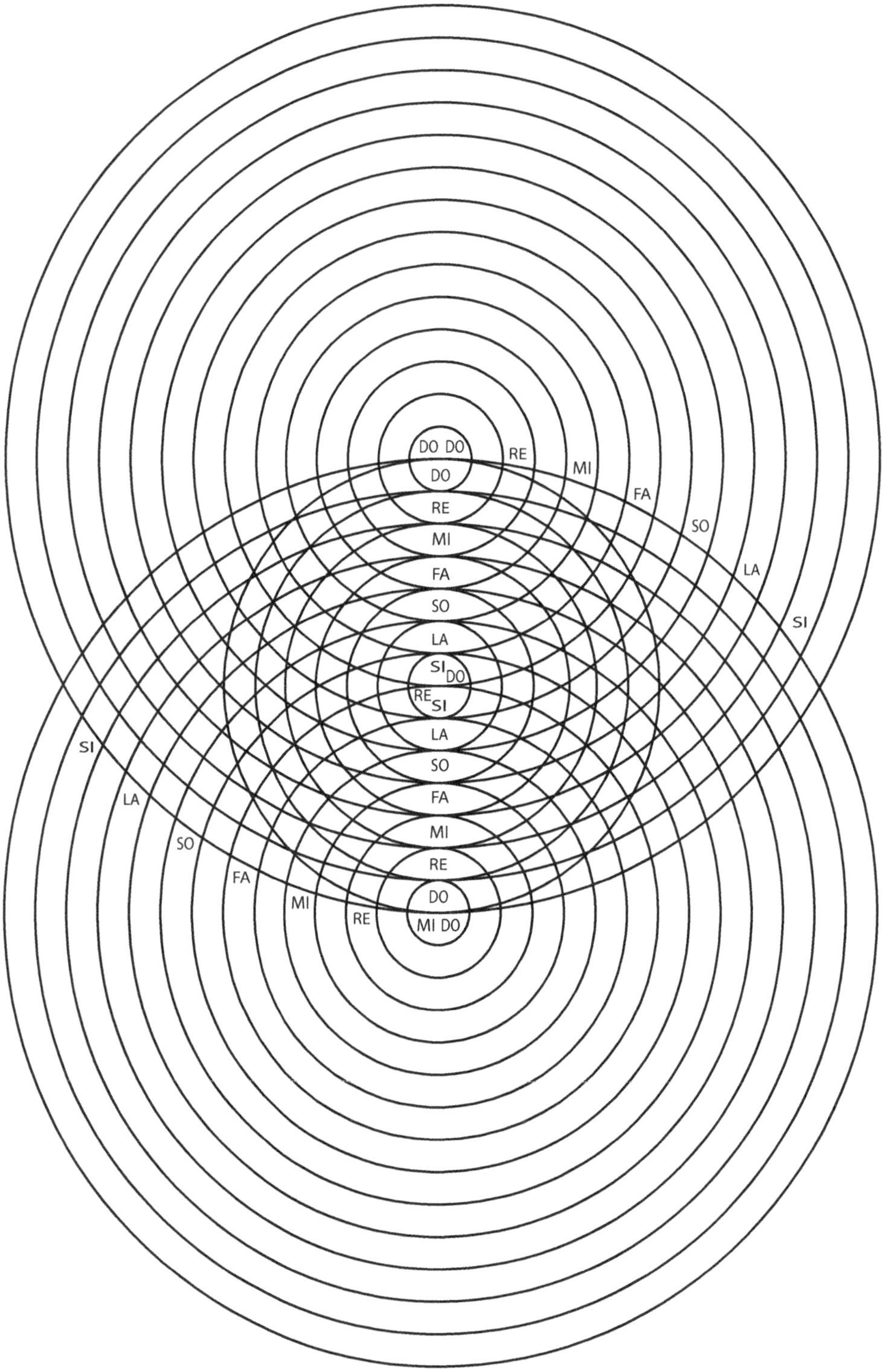

*The image above is a display of the resonant spectrum, a rainbow, of sound.*

### Figure 6. The Ordinal Sequence of the Resonant Field of Creation

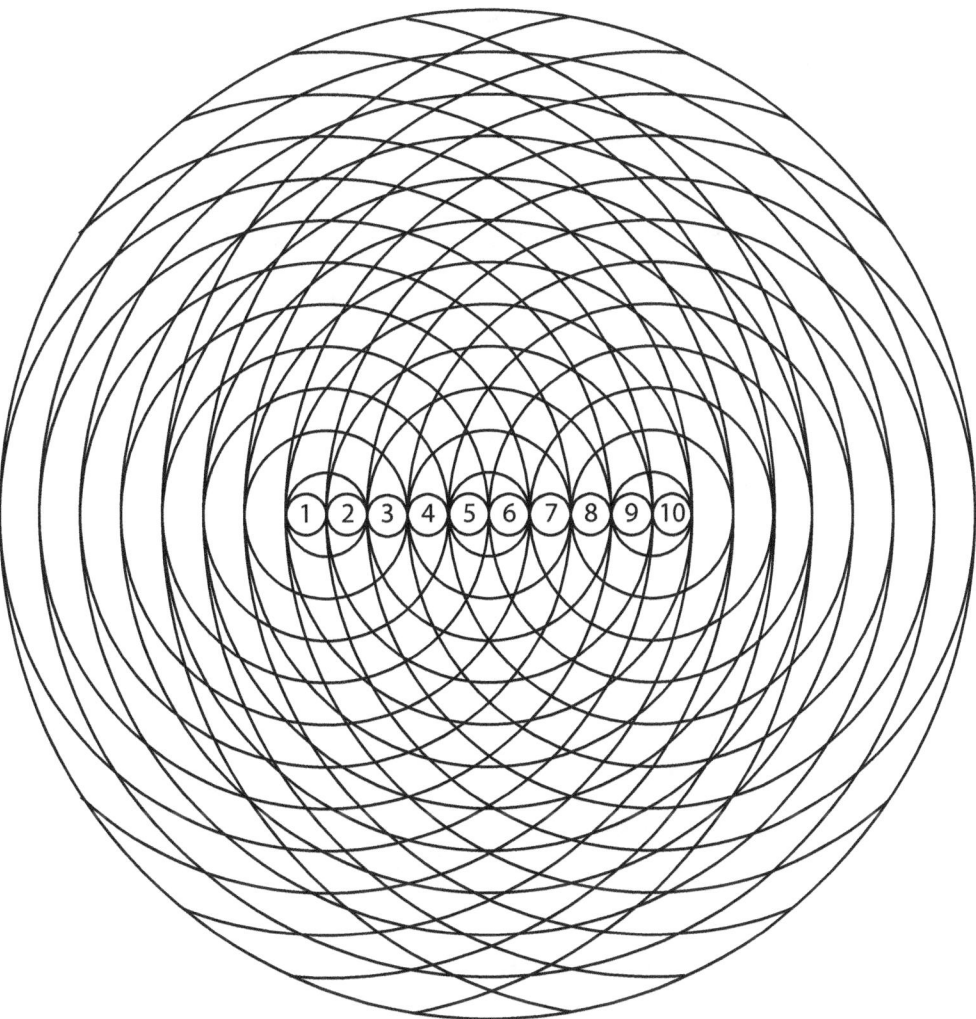

Resonance is the life force emitting and radiating from everything, creating form. Creation produces resonance. The constant vibration of the entire cosmos accumulates as a whole, forms two dimensionally with three dimensional accumulations, establishing the cosmic resonate field. Matter is the cumulative result of the Sacred Duality's and the Holy Trinity's resonant fields within the Unity of Creation. The resonant field of the cosmos, creates, shapes, and sustains structure.

Water is a liquid that has a resonant frequency becoming visible when it displaces and moves through the water in the form of waves. To demonstrate, drop a pebble into a still pool of water. Concentric waves emanate from the point of contact, where the resonating field of the water's surface was originally disturbed. Now, when the waves in the water emanate from the point of contact, are they triangles, squares, or pentads? The answer is none of the above. A pattern of circular concentric waves emanate from the point of contact. Moving outwardly rippling along the surface of the water, the concentric waves reveal a natural pattern of resonance in reaction to the pebble.

All the waves upon the oceans, lakes, and seas, are visual representations of Earth's resonant field moving along the surface of the water revealing some kind of natural rhythm, a cosmic displacement splashing down upon the shorelines, climbing up the beaches, until flowing backwards into the next wave.

# CHAPTER 2: TECTONIC RESONANT EVOLUTION

Architectural, architectonic, both define a mastery of building. The Leaning Tower of Pisa is an architectural wonder but an architectonic blunder. Architectonic is the scientific study of architecture, musical, literary or artistic structures. Architecture is the actual design and construction of the wonder.

Tectonics relates to building or construction. The forces that produce deformation in the Earth's crust, like mountains, are tectonic forces. Volcanoes are tectonic anomalies, acupressure points of our magmic world that bleeds lava, not a special kind of canoe.

The Theory of Tectonic Resonant Evolution implies that the Earth's resonant frequency is a field of energy moving through the tectonic plates of the continental land masses. Each tectonic plate and continental land mass also has a resonant frequency, a field of energy moving through all things in, near, and around the mass. Resonant frequencies move energy throughout the formation of rivers, valleys, mountains, lakes, oceans, seas, plains, deserts and forests. The formations themselves also have resonant frequencies. All of these resonating frequencies are fields of energy, moving through all things, shaping, everything smaller, the people, the animals, the trees, and all of the other critters.

Evolution is any process of growth, formation or development. In biology, evolution is the continuous genetic adaptation of organisms or species to the environment by the integrating agencies of selection, hybridization, inbreeding, and mutations. According to the definition of evolution, humanity's contiguous adaptation to our environment means we are a finer quality species today than ever previously before.

Adaptation is the core concept of evolution. If evolution is true, then every generation is getting stronger, healthier, more beautiful, and smarter. The first concept of adaptation is that all living things are in dynamic interactions with their environment. The second concept of adaptation, thus evolution, is the processes of change through natural selection.

Resonance is the third concept of adaptation. Environmental resonance shapes the continental landmasses and all the living three dimensional beings. Changes to the resonant frequencies can cause changes that mix and reshape matter. For example, teeth patterns change to fit the diet. Genetic changes are a mechanism for the evolution of new biological characteristics.

# CHAPTER 3: FLOOD ADAPTATIONS

When floods happen, the living creatures flee to higher ground and continue to survive. Receiving information through the horns from the resonant field of Earth, the horned quadrupeds can feel the floods coming long before seeing, hearing or smelling a flood. Horns are resonators that project out of the head of many creatures. Calciferous structures resonate within the resonant fields of creation, manifesting in a variety of forms amongst the creatures. Some creatures form tusks or fangs, others develop shells like crabs, snails or tortoises.

Survivors of the recurring conditions of world floods take on many forms in the animal kingdom. Giraffes have long necks so they can eat food from tree tops. The long neck of the giraffe also keeps its head above water saving it from drowning in a flood. Elephants use their trunks to breath above the surface of the water and float like hippopotami. Their fat helps keep them buoyant, surviving mass floods. Human beings most likely survived on floating debris. Species that do survive floods are the ones we see still alive.

Floods covering nearly the entire world are a regular part of the tribal stories and ancient mythologies. According to the Bible, Noah, his family and many other creatures, survived the Great Flood. Imagine a deluge, a flood, where the Ark comes to rest 18,000 feet above current sea level. Almost the entire face of the Earth was covered in water according to Noah's story.

Great floods cause the environment and genetic codes to adapt, changing their shape as a result of the drastic changes to the resonant frequencies. When the resonant frequencies change, the matter sustained by those resonant frequencies also changes. According to scripture, after the Great Flood, human beings lost approximately eighty to ninety percent of their life span.

The surface of the water after the Great Flood, is the key idea here. All the energy sweeping across the flood waters had only the mountain tops to ring, causing them to resonate at a higher pitch, resulting in gigantism. The tectonic resonance of islands and mountain tops, jutting out of the oceans, effect genetic selection during conception and birth. Perhaps the size of the Galapagos tortoise may be a clue as to the ability of tiny islands to produce the anomaly of gigantism even when there is not a flood, like the island of Sardinia.

Some of the huge megalithic structures are possibly the result of giants passing the time between meals. With limits on food supplies, giants often turned to eating the smaller people. Giants are often born with extra fingers, toes and double rows of teeth, a real dream come true for dentists! Perhaps after such global floods there is a first generation of giants afterward, and then only in several more generations as the waters recede. Giants and Flood stories go hand in hand, in folklore, myth and scripture.

The world's water level has fluctuated throughout the world's existence. There were times when the water level was hundreds of feet below where the water level is currently. Many sunken cities reside off the world's coast lines. Dwarka in India, the island of Hy Brazil off the Irish Coast, The Pyramids of Yonaguni-jima off the coast of Japan, Lion City in China and the Bimini roads to Atlantis are all cities that have been submerged from repeated flooding. During the time when the water level was low enough to support the ancient coastal cities and islands, giants existed.

Gigantism is also a result of the continental tectonic landmasses. The larger the landmass the larger the resonance of the landmass, creating larger beings. When the water levels were much lower than they are today, the landmasses were much larger too, producing giants.

# CHAPTER 4: ICE AGE ADAPTATIONS

When the world's frequency changes due to external macrocosmic influences, internal planetary influences, cycles of time, or perhaps world growth, ice ages occur. As mass amounts of water freeze, the sea and oceans recede. The ice bonds to the land, creating larger masses, changing the resonant frequencies that shape us. The reflective properties of ice create a blinding white light, heat, underground pressure, atmospheric and insular heat. How all of these might impact the genetic codes is up for debate, because the fundamental laws of tectonic resonance still needs to be applied to the equation of the theory of evolution.

The DNA of the beings surviving world catastrophes goes into a fluctuating state of emergent duress causing many distortions over time, which come and go. Mutations and changes in behavior, dwelling, breathing and heartbeat occur when the macrocosmic resonating factors change. The eyes of the creatures change reacting to the blinding white light caused by ice ages. Beings who do not die, try to recapitulate their species, causing hybrids, cross-hybrids, bottleneck speciation, and immaculate conceptions.

Norse or Teutonic myths reveal a correlation between ice and giants. The gods are the first generation of giants and the second generation killed off the first. They were all literally born from the ice. Ice sheets on top of the land masses expanding across the oceans create one larger resonating mass, resonating with every body structure, the bones, teeth, and heads, resulting in a myriad of larger-sized, larger-structured beings, with bigger bones, heads, more hair, and glandular gigantism due to the pituitary gland resonating more. Mammoths are bigger and hairier than modern elephants. The myths make more sense, realizing the bizarre, unique monsters tectonic resonance shapes. Continents coming together, drifting apart, and islands do strange things to all creatures dwelling on them. Perhaps denser bones become more dense and the thinner bones thinner.

When the ice ages melt, the movement of the glacial ice carves the rivers and mountains of the continents, bringing them together and ripping them apart, to and from whatever polar dynamic predominates. The affect that masses of ice have on this world is obvious. Ice mangles and kills just about everything the flood misses. According to geological records there are two distinctly visible charcoal layers that are almost worldwide in the northern hemisphere. Were the charcoal layers caused by fire and brimstone from above, volcanoes blowing their corks and shooting chunks of lava up into the sky, stellar plasma storms, or meteor showers and comet strikes?

The resonating field under the surface of the Earth creates and sustains the genome differently than surface dwellers. In some of the stories and myths of the survivors of world catastrophes, the people were saved, taken into underground caverns by alien looking peoples. They are indigenous, not aliens, and look like ant people, grays and reptilians. When the humans reemerged from the caverns, had they changed, mutated?

# CHAPTER 5:
# EVOLUTION OF THE HUMAN SKULL

Over time, the resonating fields that provide human existence changes. The theory of Tectonic Resonant Evolution reveals that when the resonant field changes, the shape of the human skull changes. Evolution is mutation, resulting from changes in the resonating fields, not just surviving, recapitulating, and adapting.

The objective artifacts that archeology provides, heavily favors the idea that humanity's first ancestors come from Africa. In Olduvai Gorge in Tanzania, Louis Leakey, Mary Leakey, and their son Richard Leakey found, what was once considered the oldest man-ape bones, a girl man-ape given the name Lucy.

Perhaps, when *Australopithicus* walked the Earth, Africa was much smaller at the top and larger on the bottom. Both *Australopithicus Africanus* and *Australopithicus Robustus* have really large jaws. The sagittal crest for the Robustus meant that it was more streamlined for chewing a lot of leaves, like an ape.

During Neanderthal man times, in what is now Broken Hill, Rhodesia, in Africa, 15,000-20,000 B.C., Africa must of had a little less chin, a larger back to the skull, and a really big brow ridge. Somalia looks like it was once the brow ridge, but a large chunk is now gone, and so, too, is the Neanderthal brow ridge.

We cannot forget Cro-Magnon man. Their skulls have no brow ridge and when compared to modern human skulls, the Cro-Magnon skulls are a little larger and thicker. The Cro-Magnon jawbone forms almost a 90 degree angle to the chin. Perhaps the resonating template for the Cro-Magnon jawbone is currently under water?

Why do anthropologists call us Homo sapiens instead of Hetero sapiens? Homo is Latin for man, or the earthly one. Hetero is Greek for the (other), or the 'different' one? I guess neither really fits. 'Sapiens,' doesn't mean really cool rocket man. Sapiens is derived from the word Sapient meaning wise.

According to Lloyd Pye, apes cannot modulate words or talk. Many of the anomalous skulls found are pre-human, or hominoids. Did you know Neanderthal man had a bigger brain than modern man? Maybe prehistoric men, like Cro-Magnon man, did not go extinct thirty thousand years ago. Maybe when the tectonic resonant field of Africa changes, so, too, does the human skull?

The origin of humanity intimately ties into Africa. Its tectonic resonant frequency is the source map, shaping and imprinting itself upon the DNA of humanity's ancestors. The profile of Africa is the origin for the shape of the human skull.

There are a variety of strangely shaped skulls and skeletons, like the triceratops, stegosaur, mammoths, cats, dogs, and all the other kinds. Perhaps different skull shapes are a clue as to what the landmasses may have looked like. Examining the forms of the landmass throughout history may reveal the different forms and shapes of extinct species. Antarctica looks like it has ant mandibles or like a rhino's head. Do ant people really exist? Do fairies?

When the landmass of Africa changes, so, too, does its resonant field. The resonant field of Africa creates and sustains the shape for the human skull. Changes to the human skull throughout time are a direct result of changes to the African landmass. Africa serves as a profile, the original template, for the human skull.

There are many different types of human-like skulls that exist. For example, the elongated skulls, like Akhenaton and Nefertiti. Giant skeletons exist with extra fingers and toes. Some of the giant skulls contain a jaw with double - rowed teeth. Other human-like skulls have been known to have horns. Perhaps other skulls are fitted with a hole in the forehead for a little snake-like protrusion to peer out of, like the Egyptian headdress of royalty implies in depictions. Lloyd Pye's little star-child has a uniquely shaped skull. How come archeologists haven't found any dog-men skulls as depicted in hieroglyphs?

The theory of evolution does not explain why there are so many different types of human-like skulls. There are too many mutations for the slow processes of the current theory of evolution to account for. Head-boarding, inbreeding, or hybridization are not sufficient enough explanations as to the variety of human-like skulls that do exist. The theory of tectonic resonance offers a better explanation as to why so many different types of human skulls exist. When the resonant field of the continents change or the planet changes, the DNA changes, creating a variety of human-like skulls.

There are a number of factors that might define the shape of a skull. Akhenaton, the monotheist Pharaoh of Egypt, and his wife Nefertiti were perhaps not originally from Africa. The head boarded skulls have a cranial suture where the parietal and caronal skull bone plates fuse. Akhenaton's kind, like king Tut, do not have a cranial suture which would only originate from Africa's template. Perhaps, Akhenaton, Nefertiti and king Tut were remnants of a dying breed or strain of Earth people called the Pre-Adamites, whose genetic template for their skull's shape comes from the continental resonance of South America.

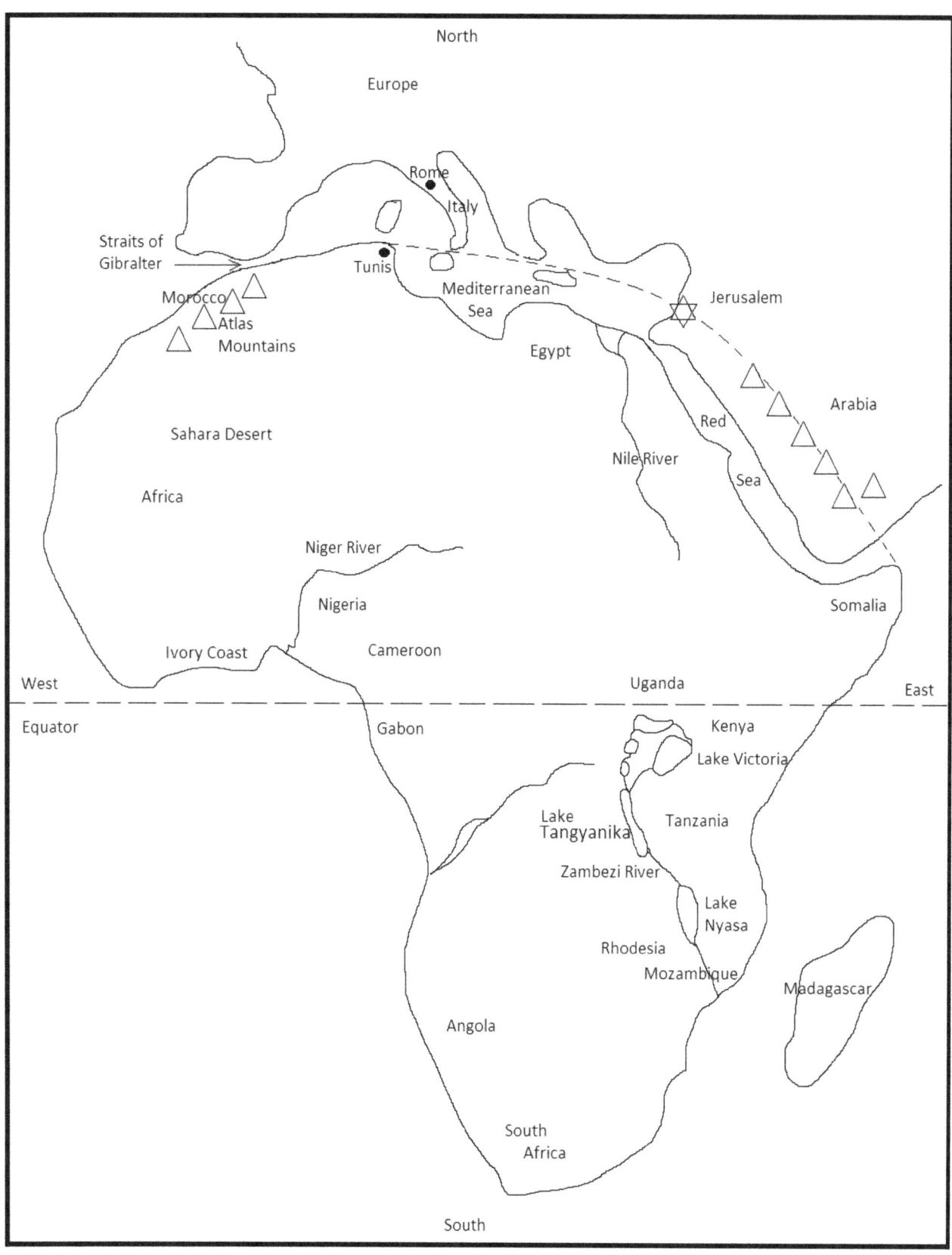

# CHAPTER 6:
# TEMPLATE FOR THE HUMAN SKULL

Science, religion, creation, knowledge of the world and self all come together in a unique eye-opening theory, the theory of Tectonic Resonant Evolution. The theory implies, as does scripture, that human beings are created in the image and reflection of our cosmic creator. The image reflected, "as above, so below," must be the continent of Africa, being either intentionally or accidentally shaped as a model template for the creation of the human skull.

There is a proportional relationship between the macrocosmic and the microcosmic principles that create and sustain the many life forms. When the macrocosmic pattern exists, like the skull of Africa, then the microcosmic pattern exists, like the human skull. In the event there are major changes to the macrocosmic template for creation then there are also major genetic changes to the microcosmic creation. Without an image, the possibility of contiguous creation may cease.

Africa looks like a human skull. Is the similarity a coincidence or proof for God? Did alien gods create humanity? Is Africa actually the remains of a macrocosmic being's skull? Perhaps God? Perhaps Goddess? There's a good chance that the vast majority of the profile of our face comes from the tectonically resonating continental landmass of Africa. Originally much larger, Africa appears to be missing several large chunks, so many I fear that Africa is the proverbial lamb that was slaughtered before the founding of the world.

As a template for the human skull, Africa, has a piece missing, east of the port city Tunis on the Mediterranean Sea. The missing piece feels like the soft spot on a baby's head, where the skull's plates fuse together. There is also an observable pulse within the soft spot of a newborn's skull. Does the human skull template, Africa, have a pulse?

One of the main figures of religion is Jesus Christ, the loving savior. Across the pulse of Africa, Rome is the center of Christianity, where the crown has a vicar and he sometimes dons a miter atop his head. Connecting to the higher world the little boot of Italy acts like an antenna, a halo, or vibrational accumulator of the resonant field of spiritual energy centering around the head. A Halo is more of a plate on a head than a head on a plate.

In the Mediterranean Sea, the word *terra* appears, meaning earth, the middle earth sea. Earth is a jumbling of the letters for Thera, the Greek goddess of healing. In ancient pronunciations the "h" is silent. The letters spelling Thera can be jumbled to spell the word heart, an organ that pumps blood and keeps us alive.

Stories of the traditions practicing resonant harmonization meditation say that the resonating spinal energy rises up through the spine and resonates with the chakras, shining. Nirvana happens when the Chakras all light up and the spinal energy releases up through the top of the head, where the soft spot is.

Another missing piece of the skull template is the frontal lobe, where the Red Sea is. The frontal lobe is the place within our heads where we watch ourselves in our dreams. If you pry open someone's eyes while they are sleeping, both are looking up toward the center. There are several major religions surrounding the Red Sea, each recognizing a prophet as the founder of their religion. The area of Jerusalem is known for its prophets having foresight into the future and the gift of prophesy, which have to come from somewhere within the matrix of the tectonic resonant field of Africa.

For human beings the northeast coastline of Somalia is the resonating template for the brow ridge, and for other creatures, the huge piece of land jutting out of the coastline forms horns. Scholars will have fun speculating upon the different types of horns that the resonating template of Africa may produce in current and extinct species.

The nose of the template is sticking out where Mozambique is having a whiff of Madagascar which may have once given human beings a larger bridge bone to their noses, and even though Madagascar is now severed from the template, people with long noses still exist. The sinus cavity is a waterway that runs along the Zambezi River up into Lake Nyasa, then up into Lake Tanganyika. Continuing along, Lake Victoria looks like the eye socket of the profile. The nations of Uganda, Kenya, Rwanda, Burundi, and Tanzania all come together around Lake Victoria. Is Lake Victoria the inspiration for the all-seeing eye of the pyramid builders?

South Africa forms the chin of the lower jaw, and the teeth. The back of the jawbone is where Angola and Namibia come together on the southwest coast of Africa. The little dent along the Angola coast continues up to the Congo River, eroding away some of the mandibular joint, where the lower jaw joins the rest of the skull. Portions of the landmass' resonant field is dampened from being underwater.

From the mouth of the Congo River to the mouth of the Niger River, where Nigeria, Cameroon, and Gabon meet looks like the ear cavity. Gabbing on, the Equator runs through the auditory and visual processing centers of the skull, deciphering the sounds and sites of the world. The Equator represents where the visible and audible spectrum of the resonant field of Earth links into the human image.

Running along the west side of Africa along the Ivory Coast all of the way around to Morocco where the Straits of Gibraltar are, is the back side of the skull-profile. On the same spot as the Straits of Gibraltar, humans have a little swirl to their hair. The shaping of our crown cowlick comes from the limit of the resonant fields of the tectonic plates of Earth coming together at the Straits of Gibraltar.

The tectonic resonance of Europe and the whole Asian land mass gently meet at the Atlas mountains. Atlas, has a relation to Atman in Hinduism, the individual self, known after enlightenment to be identical with Brahman, where all souls come from and return to. Africa, the template of the skull, may be a representation of the world in the Atlas myth. In classical Greek mythology the Titan Atlas carries the world, Africa, on his shoulder. The Atlas myth is a metaphor that means the head on your shoulders is derived from the original template for the human skull, Africa.

Jumbling the letters in the word line produces the word for Nile. Ancient people left humanity another clue as to their origins. The Nile River of the template for the human skull is a fractal pattern appearing in the same place as the fractal pattern for the suture between the front bone of the human skull and the parietal bone. Tectonic Resonant Evolution reveals Africa as a source image for the human skull.

Most shocking is the vast amount of sand covering Africa. The large amounts of sand resonate as an entire field that bakes the brains and seems a rather hellish phenomena. Jesus says, "Do not make the kingdom of heaven within you a desert." What else is there to do with a desert?

Male pattern baldness may stem from the agitation at the front and top of the profile skull. Elderly men experiencing male pattern baldness have hair growing around the back of the head, not on the top. The missing hair appears to correlate with the deserts in Africa and where the forests and grass lands grow, so, too, does the hair of old men. Nature graciously fitted elderly men with beards, to grow, trim and glue back onto their bald spot.

By day the deserts get so hot the sand itself sweats. By night the air is freezing cold. When the great Sahara releases these temperature shifts into the atmosphere they sometimes form into spiraling storms, hurricanes, which feed the Gulf Stream which some say keeps Earth's temperature stable.

If Africa is a 'God' skull, an artifact, there are probably places where distinct tribal features may arise. Skulls, or heads while alive, contain glands that produce "secretions" that are "regulatory" in essence. If the template for the human skull is the remains of a once-living being, the glands may have crystallized affecting the living beings in, near and around the crystallized glands. These crystallized glands might explain why the Pygmy are so small and other tribes like the Watussi are so much larger, almost gigantic.

The myths give us a special feeling and the theory of evolution says we are special, too. God creates humanity in His own image and that image could possibly be Africa. The skulls and faces of the many creatures, like ourselves, look fairly similar. The image of Africa as a skull profile is far too relevant to ignore.

# CHAPTER 7:
# LOBOTOMY

Ancient myths imply that there were two very strange archetypes, a western archetype and an eastern archetype. Both originate from the original tectonic resonant fields. The resonating field of North, Central and South America generate the pattern for the western archetype. Originally South America was more than double its size, including Australia and Antarctica establishing the resonant field for the elongated skull feature of the western archetype.

In Sumerian myth, the Abgal, known to the Akkadians as Apkallu, are depicted as having the body of a fish and the head of a man. Hindu myths depict the Naga as having the body of a serpent, cobra or dragon and the head of a man. The Naga with dragon bodies are associated with the dragon-people of Enlil. They have a circular fanlike flashing hood around their face. The Abgal and Naga, are the different western archetypes, surviving in image and myth world wide.

Like the South American template produces different archetypes so does the African template. In Hindu myths Shiva has four arms like her son who is born with the head of an elephant. The elephant head, much like the rhino head, probably gets its shape, or image, from the resonant field of the continental landmass of Africa. Some of the mythological beings evolving from the eastern archetype are the horned people of Enki, which include beings like Set, Satan, Loki, Lucifer and the Ba'al-ze-bubs. Kali, Sheba, Bastet, Jaguar Woman and Jezebel were all one person, as she could shape-shift into a variety of female forms. Loki in Norse mythology was also known for his shape-shifting abilities.

The eastern and western archetypes have similarities, both can shape-shift and are capable of hypnotic power over their prey. Both archetypes can live upwards of a hundred thousand years. Each archetype has horns that grow differently. The eastern archetype's horns grow on the head in pairs or sets, and the western archetype's horns grow along the spine like a dragon.

Human beings were made in the image of the Most High with no horns, reptilian characteristics, wings, fangs, tails, hooves, claws, talons, feathers, nor any evil godlike powers, just a lot of hair and a big long beard. The original resonating templates began wearing on human beings. They started conforming to the pre-existing resonate frequencies of creation becoming more like the original archetypes, the "gods". Human beings began developing different features of the gods such as horns and cloven-hooves for feet, much like Enki's, or Ba'al's people. The elongated head characteristics of Enlil's people also began taking shape in humans.

In the myths and legends, some of the first peoples practiced sacrificing their people and their children to the pre-existing eastern and western archetypes, the original gods. Mankind was literally food for the gods. Eating human beings became more like cannibalism for the gods as humans started developing characteristics like the original archetypical gods. Furious, the original archetypical gods began killing off any human god-like hybrids.

Some rare humans had developed beautiful wings and large glowing haloes of light around their heads, living thousands of years. The pre-existing archetypes known as the gods, were immortals in comparison. The original human angels began teaching humans not to worship and sacrifice their children to the gods. The food of the gods played hard to get, and the gods would rather their food simply offer itself.

Perhaps the original archetypes took human wives creating hybrids as detailed in the Bible. After man was made, man had daughters that attracted the alien gods and they took the daughters of men as wives. You know the story of Genesis, that with a few sentences covers 400,000 years of history or more. The gods all denied committing trans-sub-speciation, having offspring with a lesser species, while accusing each other of the crime.

Sumerian, Hindu, Greek and even Norse myths tell legendary tales of at least two competing archetypes of gods, super people, or aliens. The two competing archetypes are the original archetypes, eastern and western. Hindu sages speak of the two archetypes in the Mahabharata as having powerful technology, floating fortress cities, flying machines and weapons of mass destruction.

Enki struck first, striking off a chunk of South America. The chunk drifted westward and is known as Australia. In the first move, Enki reduced Enlil's archetypical super-elongated skull size by one third. The South American resonating field being much smaller without Australia attached could only sustain a smaller skull type for the Enlil people. Enlil struck back, cracking off the horn of northeast Africa, cutting the Red Sea, reducing the template of Africa to a lobotomized state. After Enlil's countermove, the tectonic resonant field of Africa could no longer sustain the image for the Enki archetype.

Enki struck again, striking off another chunk of South America. This chunk rotated 180 degrees or more. It drifted away from the tectonic template of South America towards the South Pole and is known today as Antarctica. The concussion of this act pushed out Australia even further away from South America. Enlil's second retaliation hit just east of where Tripoli is today, and knocked out a chip from the original African template, giving human beings a little soft spot on their head. Never again would the elongated head archetype or the horned archetype be able to recapitulate.

When in history, in time, did the battle take place? In my research I found the Arctic ice sheet goes back at least 50,000 years and the Antarctic ice sheet goes back to somewhere between 10,000 and 15,000 years ago. The battle must have taken place prior to the formation of the Antarctic ice sheet.

The continent of Atlantis was peopled with a third archetype, Poseidon's bat-people, the Watchers. In the convulsions somewhere between the two continents of South America and Africa, the entire artificial continent of Atlantis sunk. The caverns and hollowed tunnels filled with water drowning nearly all the inhabitants before they could fly out. Hades, or Ahriman, was one of the few survivors of the Poseidon bat archetype.

Before the Great War of Heaven took place between the original western and eastern archetypes, they lived long lives and when they died they could keep their resonant spiritual form or ghost eternal, but after the Great War of Heaven their resonant field below was so changed and damaged that neither could share in the eternal spiritual being, life above after death. The original archetypes were caste into a fate of total death for there was no hope for resurrection as a ghost eternal. All the more reason to protect the template of Humanity's arising.

After the Great War of Heaven took place between the original western and eastern archetypes, the Great Flood of Noah's time occurs wiping out nearly all the remaining inhabitants still existing after the war. Some of the western archetype survived the Great Deluge in aerial craft and a few survived in deep underground tunnels, sustained by air pockets.

Enki's archetype survived in the forms of Kukulcan and Quetzalcoatl, the old gods of the Mayan and Inca people. Inca literally comes from the name Enki. Kukulcan and Quetzalcoatl utilized the population of the Inca people to exterminate the servants of Enlil, keep watch at the entrance of Shibalba, the cavern underworld of Enlil's people, and to make sure they never come out.

In an attempt to regain the ghost eternal, the remnants of the original archetype, the horned people of Enki, continue trying to hybridize with humans with only a few successes. The hybrids were giants that humans often worked with to create sacred temples for coping with their joint needs. The giants ate humans much like the original remnant archetype. Smaller giants, like Nimrod, may have assisted humans with hunting giants into extinction as they were usually prone to falling down hard and not getting up fast enough to not be killed.

After the Biblical flood of God's wrath subsided, God crafted a new plan for humanity. Certain individuals with telepathic powers could hear God trying to speak. To anyone that could prove they were listening, God began making covenants and promises implanting the plan for our common salvation as detailed in scripture. The remnants of the original archetype, being very telepathic, heard God's plan and set themselves to the task of warring against humanity.

Salvation is the path of staying alive, whether dead or living. Heaven is here and now, whether you are alive or a spirit who once lived. There is nowhere to go when you are already there. Once you cross that bridge you are already on the other side.

# CHAPTER 8:
## POSEIDON

As a little fellow I remember someone handing me a conch shell saying. "Put it to your ear and you can hear the oceans." Feeling a bit childish, I put the conch shell to my ear and guess what I heard. "Run away with the conch shell?" No. Did I hear, "Hello? Is anyone there? This is Poseidon calling. To whom am I speaking?" No, the sound I heard from the conch shell is how I imagine outer space or the deepest depths of the oceans sounding. Next time that you get a chance to, check out the difference between the interior and exterior resonances of definitive variation. Hold your hands tight against your ears and say, "Om." Then, remove your hands from your ears and say, "Om." There is a tremendous difference in the way it sounds.

Myth details the founder of Atlantis as Poseidon, a self-appointed king, an alien, the god of the seas and oceans. The Greek god Poseidon probably didn't change his name to Neptune nor turn his trident into chewing gum. Poseidon's archetype, the bat people, have facial shape-shifting abilities and talons for feet like an eagle, almost never leaving a single print. Much like bats, bat people have really big sonar sensing ears and often have difficulty seeing in daylight.

Poseidon took a human wife and had five sets of twins with her. Some of their children also took human wives, producing a hybrid race known as the mighty men of old, the giants. The Sumerian Kings list details these giants as the first kings, reigning for thousands of years each. In the book of Enoch, Poseidon's people are known as the watchers, the sons of Heaven, or the fallen ones, and their hybrid children, the Nephilim, were giants.

Poseidon's symbol is the trident. It looks like a devil's pitch fork and sounds like a tuning fork. The top part of the trident looks like the Hebrew symbol Shin, given a numerical value of seventy-two, and looks like the central three candle holders for the golden lampstand. The golden lampstand is central to Christian, Islam, Hebrew and many faiths. If you take the map of Atlantis as the concentric rings, cross-sect it with the golden lampstand, the obvious reveals itself, the golden lampstand is one-half, the lower half of the map of Atlantis.

**Figure 7a.
Poseidon's Trident**

Poseidon's trident sounds like a tuning fork, resonance, and resembles the first three candle holders of the Golden Lampstand.

**Figure 7b.
The Golden Lampstand**

The Golden Lampstand completes half the map of Atlantis.

**Figure 7c.
The Map of Atlantis**

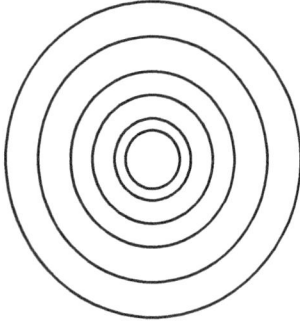

The map of Atlantis as depicted by Plato.

**Figure 7d.
Rainbow of Resonance**

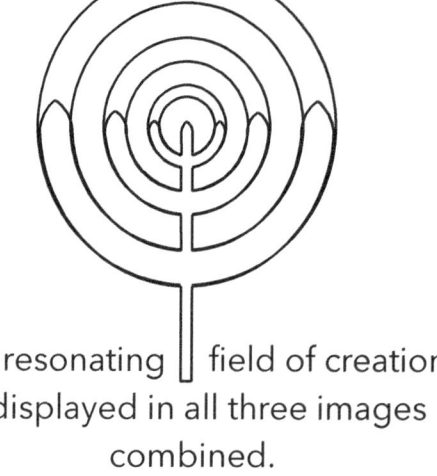

The resonating field of creation is displayed in all three images combined.

Plato's depiction of Atlantis is famous. He received the description from his great Uncle Solon, whose description of Atlantis came from an unknown ancient order of initiates in Egypt. The map of Atlantis was possibly the ancient symbol for Atlantis. The rings of Atlantis most likely bobbed up and down in the water like a wave maker. Plato's description is of a tectonically resonating mass, intentionally designed in concentric rings for receiving the tectonic vibration of the ocean to keep afloat.

After Atlantis sunk the story continues in Central America. In the Popol Vuh of the Mayans, the bat-god Camazotz, forced the human population to provide blood sacrifices and offerings. The name Camazaotz translates as "death bat" or "snatch-bat." What defeats or keeps Camazotz away is the light of day, the Sun, as bat people are, as is said, "Blind as bats." Over time the original stories like, the Great War in Heaven, the myth of Poseidon, the Mayan myths and others stories get mutated and mixed together incorporating a much bigger confusing picture, laying a blizzard of snow covering the fresh hoof prints of the world's earliest histories.

# CHAPTER 9:
# THE SACRED TORTOISE

From the very beginning mankind has been playing with the one thing slow enough for him to catch, a tortoise. It is a schemata for our puzzled enlightenment. Quietly munching on the strawberries in the garden the tortoise says, "Catch me, if you can?"

Catching the tortoise, ancient peoples create language, discover geometry, and while deciphering the geometry, invent the protractor, a tool for measuring Earth's dome. Patterns of geometric resonance construct the tortoise, a living reflection of Earth's resonant field. A comparison in the geometry between the Egyptian tortoise and the South-African tortoise reveals the interrelationship of the original tree of lights, highlighting the original thirteen sefiroth in the ancient alpha numeric system modern mystics call the Kaballah, or Cabala.

The geometric pattern of the firmament, Heavens, Dome, tent of the Most High, is on the back of our friend the tortoise, a time piece of creation. Along the center of the spine ridge of the tortoise aligns three hexagons surrounded by ten pentagons, which all are surrounded by twenty-four and a third squares. Each square represents an hour. The world's resonant frequencies displace themselves, carving the segmented geometry of the sacred tortoise.

Ptah is the name of the first creator god of the Nile Valley peoples, the protector of the Nile, originally the Hapi or Hapi River. The culture and knowledge of the Nile Valley people first centered around the recognition of the tortoises as sacred beings. Building tortoise effigies all around the world, the Nile Valley people spread their knowledge of the tortoise, a totem of the Great White Goddess, Tara.

East of the Great Pyramid sits the Sphinx of Egypt, originally a tortoise effigy named Ptah. The 'P' and 'h' are silent. The sound 'Ta' is also the sound a tortoise makes as it retreats into its shell. Studying the tortoise the ancients derive the sounds and symbols for letters. Ta or Ptah, is the sound of the breath of life, of the creator turtle, like 'Om' is for the Buddhist. The Ta sound breathes life into the words of the oracular Tarot Card system. This system originates from Egypt where the letter T is a small class t, a cross, for the four directions, the four winds that blow the breath of life into the living things, and creating the sounds they make.

Noah's Ark is synonymous with the Ark of Adam Cadmon, the sarcophagus of the First Man, which comes to rest upon Mount Ararat. The backwards spelling of Ararat is Tarara, the name of the original tortoise effigy east of the Great Pyramid. Peoples west of the Hapi River call their creator god 'Ptah-hra-ra.' East of the Hapi the peoples call their creator god 'Har-rha-aht.' Both peoples east and west of the Hapi river pronounce the name the same way, as Ararat.

As a symbol, the silent 'H' depicts the sacred marriage of Heaven and Earth. The dividing middle line of the letter H is a symbol for the division in the world, like the equator, or like Heaven and Hell. Dropping from many words throughout time, like Har-meggido becoming Armageddon, the silent H is lost in the written translations of oral traditions.

Early in Egyptian civilization the turtle effigies were destroyed and replaced with a scarab named Ptah. This marked a change from the flat stationary turtle world view into a round rolling ball of dung world view. As changes occur throughout time, the Nile Valley people lose their knowledge of reality, the sacred turtle, language, geometry, and their symbolic first god and goddess.

The original alchemical intoning of the Holy Name of the Most High, also becomes lost. According to legend, when spoken, the Holy Name of the Most High resonates throughout the lands providing annual fertility. In the Bible, when the Most High speaks, the words resonate manifesting reality. Resonance is the key to creation, recreation, and healing.

The letter I is pronounced E, in the ancient oracle book of ancient China, the 'I Ching', and is also known as the Book of Changes. A proper oracle literally speaks. The I Ching responds to any question asked in the form of a hexagram. Hexagrams are composed of six lines, open and closed.

Hexagram 27 of the 'I Ching' represents the letter I, titled nourishment, is literally translated as "jaws' not "nourishment." Looking at hexagram 27 the bottom three lines compose the bottom jaw, and the top three lines compose the top jaw. The "I" teeth are the first fourth teeth we are born with and the "eye" teeth are the four canine teeth.

**Figure 8a.
Hexagram 27**

Nourishment, "Jaws"

**Figure 8c.
Mouth**

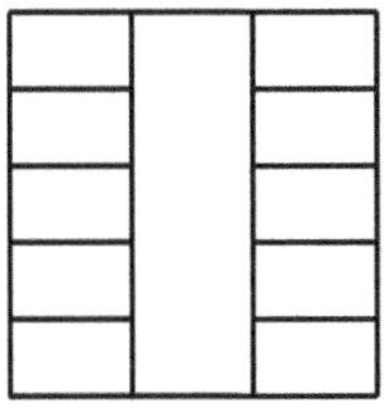

The space in the middle of the hexagram looks like an "I" or an open mouth.

**Figure 8b.
The Eye Teeth**

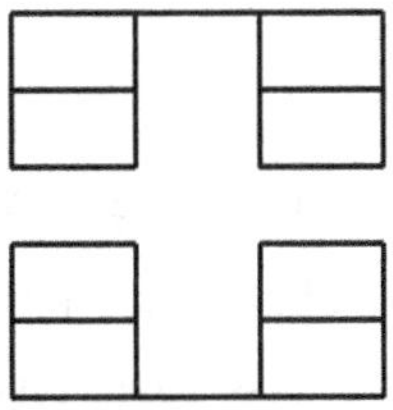

In dentistry the first four teeth people are born with are called their "I" teeth. The Canine teeth are also known as the eye teeth and are the next teeth before the Bicuspids.

**Figure 8d. Nourishing the Mouth**

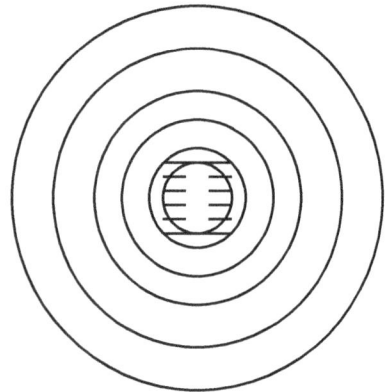

**Figure 8e. Jaws**

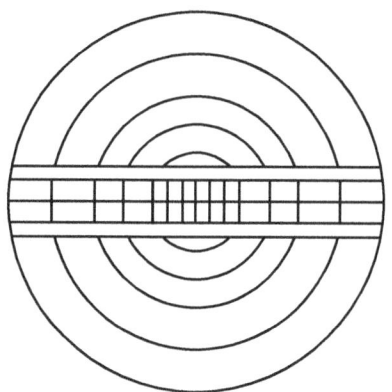

*Hexagram 27 fits inside the Map of Atlantis, extruding the lines across the resonant field provides a clue as to how the teeth are formed.*

Each human beings jaw is composed of a set of eight named teeth, composing the sixteen teeth for both the bottom and upper jaw, adding up to a total of thirty-two permanent teeth. The teeth, the jaw, fits into the Map of Atlantis. Beginning with the first three named teeth, the two incisors, and the eye teeth are composed within the first two dimensions given in Plato's description of Atlantis. The inner circle's diameter is five and the width of the next circle is one, adding up to six. Each of the first three named teeth all have one root and are the first six teeth for the upper and lower jaw, composing the first twelve teeth.

The next two circles in Plato's description of Atlantis are given the width of two. After the eye teeth comes the premolars or bicuspids. There are only two bicuspids each having two roots. The dimensions of the last two outer circles given by Plato have a width of three. Human beings have three sets of molars, all of which have three roots. Adding Plato's dimension of Atlantis together equals sixteen the total number of teeth for the upper and lower jaw, adding together for a total of thirty-two teeth. In the I-Ching there are thirty-two yin hexagrams and thirty-two yang hexagrams for a total of sixty-four hexagrams, fully complimenting the duality of the feminine and masculine.

Plato's Map of Atlantis originates from an unknown ancient Egyptian order. Ancient orders that kept secret knowledge hidden are also called esoteric circles. In order for the Map of Atlantis to become the official Seal of Atlantis another circle must be drawn around the entire map thus providing the additional circle for the wisdom teeth. A seal, like a mandala, is an image contained within a circle.

Letters and symbols were first derived to help traditional oral story tellers remember the story. The letters H and I are derived from the cross-section of the underside of the tortoise where its resonant field is centered in the first, second and third dimensions. Combining the letters H and I creates the four central squares of the chessboard, and is the first symbol for the "four quarters" of the World, and the four directions.

**Figure 9.
Four Quarters of the World**

And the message from under the World, as a symbolic representation of the tortoise, is Hi.

Sometimes we get just enough information to sustain our interest longer than an average amount of time. Mainstream seekers know about the twelve apostles and think Jesus' favorite number is twelve or one apostle for each house in the astrological zodiac. The real number is thirteen, including their founding member Jesus. The ancients derive the thirteen segments for the universal system of cosmic knowledge from the top shell of the terrapin. Its original spelling is Terraphim. They swim, walk, float and according to some stories, fly. Hebrew mystics consider the Terraphim to be one of several classes of angels.

In ancient times peoples record witnessing machines that look like flying tortoises. There are ten pairs of flying tortoises as the author of the I Ching depicts. In the Hexagram Sun/Decrease the fifth line says: "Someone does indeed increase him. Ten pairs of tortoises cannot oppose it. Supreme good fortune." The explanation of the line says: "that all oracles including those read from the shells of tortoises, concur in giving favorable signs, luck is ordained from on high." This is probably the luckiest line of the whole I Ching.

The I Ching is an oracle that the author derives from the sacred geometry of a turtle. Hexagrams 64 and 63 disclose a war with non-human entities, a war between devils, gods, and men. Both hexagrams speak of the Devil's country as if it is everywhere, and that there is a war between the gods and the demons that is raging, and in the end, the outcome will decide the fate and destiny of the human race. Jesus tells us who the devil and his minions are. The Devil is in the details whenever a flowering civilization begins to self destruct.

Humanity's knowledge discerning cosmic universal laws from the tortoise instills a godlike insight into the hierarchy of geo-symmetrical consciousness. The original archetypical gods, the horned people of Enki viewed humanity's expanse of knowledge as threatening. Through war, chaos and manipulation of the people, the original archetypical gods destroyed many of humanity's greatest civilizations.

The Churning of the Sea myth tells a tale of a battle over the world, between the Devas and Asuras. These races battle over the world tree on Mount Mandara, growing out of Vishnu. Vishnu is in the form of the gigantic sacred tortoise and is being used as a pivot to churn the sea and make ambrosia for the gods. There are mythological beings, the Asuras, who are white, red and black, with horns on top, very devilish looking. They are playing tug of war, with Vasuki, a snake, as the rope, wrapping around the world tree. On the other team the Devas are composed of more human-like people, some with pale blue-gray skin. The Devas and Asuras were considered the gods. The artistic rendition of the Churning of the Sea is a depiction of a battle between the Asuras, or devils, and earlier god-men over the control of the world, and over control of the gift of sacred geometry provided by the sacred tortoise.

Authors of historical myth tell stories of sacred trees so big they held up the very sky. In the myths people climb the colossal sacred trees high into the heavens, above the sky. Looking out from the sacred trees' tops people could see the ends of the world and converse with the Gods. The mega-trees dwarfing others were a miracle but drank so much water that little else could grow. Sucking all the water out of the ground, the environment around the colossal trees turned into deserts.

Early terra formers probably cut down the colossal trees, perhaps utilizing the celestial logs of creation to build the resonating pillars of the four corners of the world, on which the Hemispheric Firmamental Stationary Flat Earth stands. Interfering with Earth's resonant field, stumps of the great trees draw electricity into themselves causing petrification. Many of the mesas like Devil's Tower, and mountains like Mount Meru and Everest may be stumps of the oldest trees. Pieces of a primeval reality exist in the mega-stumps and ancient myths of the world.

# CHAPTER 10:
# TEMPLE BUILDERS

When viewing a stela of ancient Egyptian art pay special attention as the right eye's view is on the left side and the left eye's view is on the right side. The two one-eyed jacks, from common playing cards, are an old throwback of the old story about two brothers, the Seraphim and Cherubim. They have a predisposition to be right or left side dominant resulting from the original resonating templates, South America and Africa.

In the Osiris myth, Set kills Osiris, cutting him up like a checker or chessboard, seven times vertically and seven more times horizontally. The story depicts Seven Foldness. To better understand Seven Foldness start by folding a square piece of paper in half, from left to right, forming a rectangle. The second and third steps are in the same direction, folding the rectangle in half with each step. Now fold the rectangle from top to bottom in half three times to form a smaller square. Seven Foldness is one of many methods used to graph myth and its variations.

Lots of ancient myth has been scribed on scrolls for future generations. Some scrolls were designed more like lamp shades, hollow, with crafted images etched through the surface of the cylindrical shape. Ancient oral story tellers would hold the scroll up by the fire light. The light casting through the crafted images paints a picture of myths and legends upon the surface of the cave. So, if you want to find out where Osiris's phallus wound up, how he gets trapped in a tree, or how his head fits into the pyramid, you will have to find the right stela for the right myth, copy it, make the folds or cuts and put the story together, piece by piece.

The symbol of the All-Seeing Eye on the dollar bill, sitting atop a pyramid of thirteen levels comes from an ancient story, predating Atlantis and human history, a legend that no record of currently exists. No one ever speaks of the legend of the all-seeing eye as no one has lived long enough to hear the story.

Why just one eye? What happened to the other eye? What is the eye looking at? Is it a right eye or a left eye, or maybe the eye of the Cyclops? Or perhaps our inner I? Is it the eye of Osiris or Thoth? Or the eye of Horus, Isis, or whomever the overseers call themselves. Maybe Ymir, or even Odin? Africa is the resonant continental template, the profile for the skull and the eye is Lake Victoria. The all-seeing eye is telling us to: "Watch where we are going."

Speaking of fragments of ancient knowledge; Noah's grandson, Mizraim who is also known as Mithras, Mitra, Maitreya, names that are different dialectic versions of the same person, is the father of the ancient brotherhood of seven. They are responsible for passing down sacred knowledge for future generations and utilizing that knowledge to build the Sacred Temples and Sites existing around the world.

Mitra learned everything he knows about temple building from his grandfather Noah, who built the ark that survived the Great Deluge. The Ark is built on the same principles guiding the construction of many of the great holy sacred sites around the world, sacred geometry. In an effort to pass that knowledge of the unsinkable ship down for future generations, Mizriam wrote down the Dimensions of the ship as depicted in Genesis, Chapter 6, verses 15-16. Noah's Ark fits exactly like Jonah in the whale, perfectly, inside of the Vesica Pisces of the blended duality, in length and width, exactly inside the limits of the resonant field's containment potential. The top of the Ark is designed very close to the water level, so that the waves, no matter how large, rolled over the Ark without rolling the Ark over.

### Figure 10a. Top View. Noah's Ark.

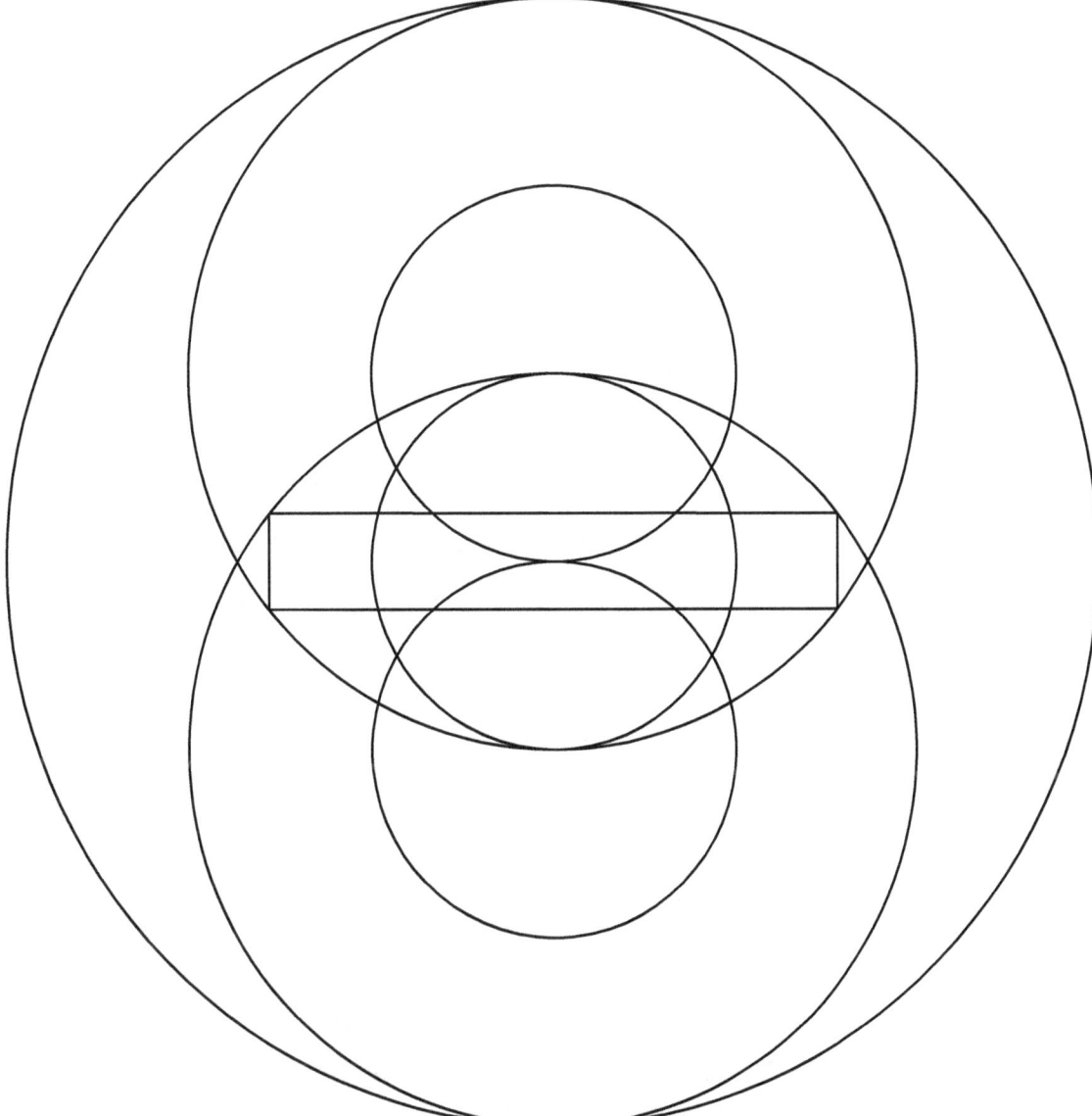

*The horizontal lines are the length of 300 cubits and vertical lines are the width of 50 cubits. Genesis 6:15 says, "And this is how you will make it: three hundred cubit's the length of the ark, fifty cubits its width, and thirty cubits it's height."*

### Figure 10b. Side View. Noah's Ark.

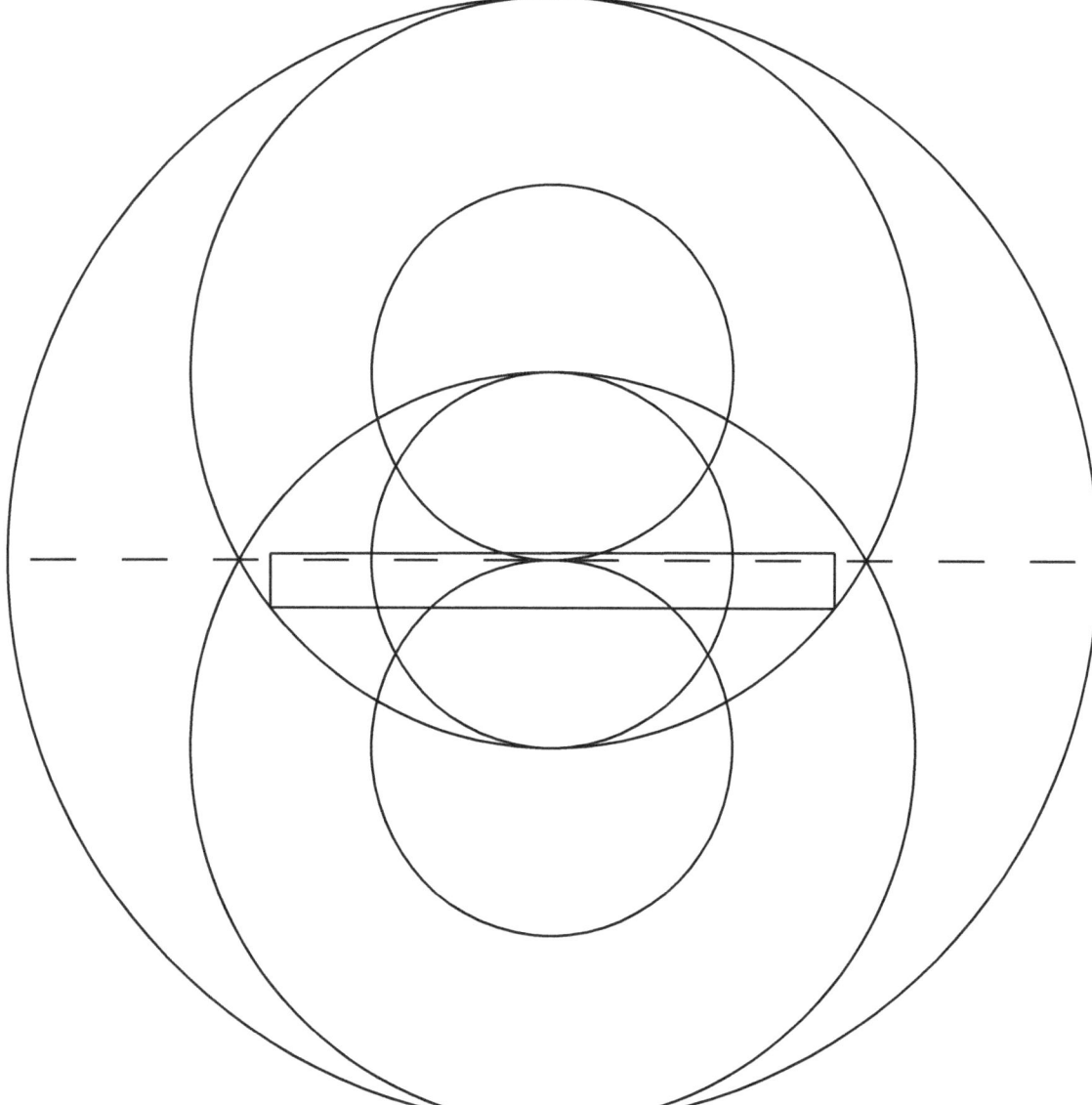

*According to Genesis 6:15 the height of the ark is 30 cubits. The vertical lines depict 30 cubits in height and the horizontal lines 300 cubits in length. Noah's Ark is just above the dashed line of the water level.*

Sphinxes are ancient creatures of mythological lore. The Great Sphinx of Egypt is not the only sphinx in the world to be carved by ancient peoples. Under the guardian sphinx's paw at the Temple of Heaven in ancient China is a three-dimensional model of the spheroid, inspired by the flower of life mandala derived from the three hexagons topping the top shell of our tortoise.

The ancient brotherhood of seven left their seal, the Flower of Life mandala, on the temples they assisted in designing and building. The flower of life looks like a flower, a lotus for example. Biblical tales tell us Mizraim moved from Shinar to the land of baked bricks, Africa, where Mizraim and his followers built the pyramids. Mizraim had seven sons, the seven sages, the Aka, who built the Great Pyramid according to the Edfu text.

Mitra and the brotherhood of seven developed a technique for moving the large building blocks of the holy sacred temples by inventing the miter. A miter can be used as a wedge or a fulcrum. The leverage provided by the miter is utilized to lift the large temples stones up enough to place stone balls under the temple blocks for them to roll on. Mitra and the brotherhood could roll the colossal temple blocks in any direction. To stop the blocks, miters are used like wedges to stop the balls from rolling. Knowledge of resonance is not lost to Mitra and he may have used conic resonators, instrumental horns, or banged gongs to match the frequency of the temple blocks, greatly reducing their weight, causing them to levitate slightly moving with ease.

Under the supervision of Osiris, the Osirions, turned the original turtle effigy into Anubis, a jackal head, who weighs the hearts of the dead. After the Great Flood, as the authors of the Bible record, Noah's grandson, Mizraim discovers the Anubis effigy in ruins, and dedicates a new effigy, the leonine Sphinx of Egypt. However, when the island of Santorini exploded the Giza plateau was deluged covering the Sphinx. Then, Khufu finds the Sphinx and changes the lion's face into a man's face with a cobra headdress.

The original ancestors working with a technology similar to seismographics, placed pyramids all over the world to prevent earthquakes, and to stop the disintegration of the continental landmasses. Pyramids prevent the continent of Africa from colliding into Eurasia and stabilize the thinning of Central America. The Great Pyramid stabilizes the human skull template, which may be saving us.

Earth's tectonic resonating masses change over time. When the masses of the Earth change so do the peoples building the temples and holy sacred sites around the world. After each passing wave of destruction, floods, wars and ice ages, the peoples build new monuments, sometimes over old monuments leaving clues for generations to follow.

Many of the earliest holy sacred temples are resonating chambers. Intoning holy vowel sounds monotonal, diatonal or tritonal voices and instruments, repeating their vocalizations and chantings, ancient people communicate with their ancestors. With practice, ancient groups could open portals, walk in and out of layers of time, and see one hundred, possibly thousands of years into the future or past. Singing, chanting, or intoning a mantra, inside the holy sacred temples excites the body's chakras, evoking the sensation of spiritualization of the spirit body, the astral body, the physical body, and the soul. The experience is healing, pleasing, and often referred to as enlightenment, illumination, nirvana or even ecstasy.

The sacred canticles and the magical harps of angels plucking in sacred sequence resonate the firmament, much like the monks in their temples. Ancient architecture mirrors the above bringing the image of Heaven to Earth. Resonant harmony is key and seems to be applied to all temple buildings. Feng Shui is an ancient Oriental science of healing by the careful placement of the right things in the right places. Things like architecture and landscape have a profound affect upon our common health.

# CHAPTER 11:
## GEODESY

Here is a big secret: The Earth has secret ley lines where mysterious powerful phenomena happens. Many of the sacred sites and temples, like the Rose Line Chapel in France, exist along the mysterious secret ley lines of the Earth. Geodesy is the science of measuring the curvature of the Earth to determine the Earth's shape. Is there a connection between Earth's encircling gridlines and the holy sacred temples?

However, that is not the big secret. The imaginary latitude and longitude lines paint a picture of the resonant field grid containing Earth. Is the resonant field a power grid or the blueprint that lends itself to the creation of our resonant structures, like our bones? The answers are like whispers in the ear, listen closely.

The big secret is that, the Equator, the biggest of all of these imaginary lines of latitude, is where the greatest amount of the world's resonance accumulates upon the surface of the Earth. The Equator is given a value of zero, degrees and it whispers directly into the ear of the template for the profile of the skull, "Rise and Shine!" Exactly halfway between the polar limits of the tectonic resonant field, between 19.5 degrees above and below the Equator, Earth's volcanoes and the Sun's spots seem to have the greatest amount of resonant behavior.

Maps using latitude and longitude go way back in history, to the time of the Greco-Egyptian astronomer Claudius Ptolemy. There are *portolani*, or seamen's charts which are thought to be generally accurate, although at that time there were no systems, or standards for the measuring of great distances. Sailing the world, maritime people provide no documentation of a globe Earth.

The Babylonian, Sumerian, and other ancient civilizations probably should have discovered the curvature of the Earth or at least theorized about the Earth's shape in the Academy. One of the oldest sciences is astronomy which saw the sky as a curved dome. The creation myths say the Earth is a stationary circular disc in a dome, with rivers, mountains, seas, deserts, oceans, and continents. For a long time the peoples of Earth believed the Earth was flat, round and with a dome on top like a snow globe.

Pythagoras and the Greek philosopher Aristotle were the first to postulate a spherical Earth. Aristotle gives an estimate for the Earth's circumference at 400,000 stadia. The word stadia probably comes from the word stadion. Stadion is the name of the running game for the Olympic games in Attica, measuring approximately 200 yards. Eratosthenes later derives what he believes is a more accurate measurement for the circumference of the Earth at approximately 250,000 stadia. The word circumference by definition is the distance covering the outside of a circle, not a sphere. If the Earth is a sphere the correct term is the limits of the Earth's spheroidicity.

There are no more refinements to the spheroid Earth model theory of Erastosthenes until the oblate spheroid Earth comes into question in the late 17$^{th}$ and early 18$^{th}$ centuries. Utilizing a method of triangulation Wille brord Snell, (1591-1626) estimates the Earth's circumference to be about 38,653km. When Isaac Newton tries to prove the Earth attracts the Moon as a principle governing force he uses Picard's (1668) measurement of angles of triangulation. A new age of geodesy begins to move away from the spherical Earth to an assumption of the Copernican system of axial rotation giving the Earth's form that of an oblate spheroid.

The Van Allen belt is an invisible planetary force field of magnetic lines arching out into space protecting the Earth from the Sun's lethal high energy particles moving at millions of miles an hour. Some of these cosmic rays from the Sun enter in to Earth's upper atmospheres, striking the oxygen and nitrogen particles, producing the aurora borealis and aurora australis. Rock samples reveal the Earth's magnetic pole flips every few hundred thousand years. When the magnetic poles of Earth flip, does the magnetic field of Earth disengage and do all the landmasses begin floating, like a magic carpet ride?

From humanity's perspective the universe is a big place. The ancients agree, clarifying Earth as the center of the universe. All laws relevant to centricity and harmony of the cosmos are in perfect equipoise at the center, making Earth the virtual pivot point of Creation and possibly the most miraculous place or point in the universe.

Detecting all of the stellar formations in the field of the cosmic background radiation of the universe's resonant field, the COBE Satellite paints a picture of Creation which was revealed in the 1990's. The data gathered was reentered and then exponentially reentered again and again. The picture changed into one where creation is a mega galaxy of galaxies, and at the center of the picture is Earth.

Each star emits its own resonant frequency in the form of sound and light. The distance between the stars is the self-sustaining aspect of the macrocosm, the universe at large. There must be a whole lot of sound coming from the stars all centering in and around the Milky Way. How can sound or colors exist on Earth if Earth is surrounded by the vacuum of space? There must be a pressurized dome of some kind or Earth would be nothing to speak of, hear or see. Very simply there would be no life, rain, and the oceans would drift off with the next wave. More noise resonating in the dome can cause the tectonic plates to tremble disharmoniously, disintegrating faster, shortening humanity's life span, destroying the very world of which we are still learning.

# CHAPTER 12:
## THE LIVING EARTH

Aside from tectonic resonance, the other big part of the resonator of Earth is the container system itself, as an outer aspect, shell or skin. Is the Earth a living being? Perhaps our world is a living, breathing, resonating orb. Earth's outer shell serves as an accumulator of vibrations, which circulates through the tectonic plating, in the inner shell, as one single cosmic breath of life is inhaled through the center of the Earth, from the North Pole. The North Pole acts as a mouth for the nourishment of Earth as a living organism.

Earth's North Pole is a hole that opens and closes once a year, a breath of life that sustains and regenerates the Earth's essence by exercising its outer skin. The breathing action of the Earth vibrates the torus field of Earth's magnetic plate, gathering a constant supply of electrical energy from the outside of Earth's skin, circulating the electrical energy back down into the gullet of the hemisphere, the magnetic North or mouth of Earth.

On the bottom of Earth's magnetic plate is the South Pole, a hole that opens and discharges the breath taken in through the North Pole. The circulating magnetically charged electrical energies are exhaled in a contraction, pushing down, and out, through the bottom of the Earth's outer skin.

Resting for an interval, about halfway, thus level with the magnetic plate, the next breath stretches the Earth's outer skin up, over, and back down the vortex feeding into the North Pole. In the interval, electrical energy, food, and nourishment, gathers on the exterior of the Earth's outer skin, circulating back into Earth.

Really, there is no South Pole, and there is no North Pole. Earth is a living orb, with a magnetically charging plate. The action of the Sun and the Moon charge the Earth's plate and instigates the breathing process. The living orb, Earth, maintains a frequency throughout the entire process of charging and discharging itself. This occurs annually.

Perhaps, the active and constant movement of the Sun and the Moon, of "a day and a night," as in the tale of Atlantis, or in the first days and nights of creation in Genesis, is a common clue that the ancients recognize the importance of the repetitious cycles of the Sun and the Moon. They are the life-giving, life-sustaining duality, that provides life's heartbeat a rhythm, like sunrise, sunset, day and night. Or is it sunrise moonrise, maybe sunset moonset?

In the Good Book, The Bible, Genesis, Chapter 1, verse 2: "Now the earth proved to be formless and waste and there was darkness upon the SUR-FACE of the watery deep; and God's spirit, (or active force), (what, did he die?), was moving to and fro over the SUR-FACE of the waters." The words "to and fro" sounds like resonant modulation. The prefix Sur- means super and before. The word surface literally means, 'before super' face.

In the tectonic resonance theory, the sustaining force, the "to and fro," is a frequency constantly vibrating throughout the world in a cycle of circulation. What modulates the sustaining force? Is there a repeating rhythm, a flow of regular movement, like walking left, right, left, right?

The temperature exchange between the Sun and the Moon plays an important role. When the temperature rises the tectonic plates charge, expand, and rise by day, by night, contracts, shrinks, and sinks. Day and night are a single natural hover, a breath, a world motion of a resonant fluctuation, like the heartbeat of the Creator Earth, in cohesion with the Creator Sun and Moon. Life springs back to life in Spring when the weather is warmer and into full foliage in Summer. In the Winter everything recedes, withdrawals, or hibernates. The tectonic plates also rest and contract during the winter while the weather is coolest. The entire year is a cosmic breath of life, but, more like an annual meal.

One of the many aspects of self-knowledge is an understanding of oneself as a cosmic metaphor. If the Earth breathes once a year, then a one-hundred-year-old human has lived for one hundred Earth seconds, nearly two whole minutes. Life feels like a strangely brief acquaintance.

Earth and the stars are all resonators, living beings of the megalocosmos. Their resonance creates and sustains existence. Like all living beings Earth eats and grows.

Stellar material is food, plunging into the belly of Earth. To avoid being struck by stellar material too large to eat, the Earth will move, reacting to its environment. When the Earth moves, it takes a deep breath in, pulling down lots of water and cold air through the North Pole. The primary cause for periodic mass destruction upon Earth, is the Earth itself.

# CHAPTER 13: BEING A RESONANT BEING

Time passes. The image changes little. Humans arise from the very image, like a photo coming to life, the clay model breathes in the living breath of life. Imagine being someone else in history for a moment, surviving world catastrophes, wars, famine, disease, performing basic functions like getting the water, chopping the wood, feeding the fire and family, going to work, and with any remaining time scribing sacred knowledge for future generations. Life is time and time is life. The sources of time provides a recurrent and stable rhythm, cycling constantly into the resonating fields of all other resonating sources.

Resonators like the Sun, Earth, and Moon, have intersecting fields of resonance where they combine, creating the possibility for one, two, and three-dimensional beings, like humans and other life forms. Human beings are a direct living manifestation of reality. The salivating truth is human beings swallow water in one swallow per second, their hearts beat once every second and with every other second breathes. Tick-tock goes the clock, tick-tock goes the human being.

Resonance is a phenomenon that occurs when there are two or more objects having the same natural frequency of vibration. Human beings bodies have just about two of everything. Does this mean human beings are a resonance pattern come to life? Everything is a resonance pattern. Think about it! The resonant frequencies creating humanity provide human beings with physical structure, each having their own resonant frequency, as seen in the microscopic image of our DNA.

Parity is the study of the predisposition to be born left-handed, right-handed or ambidextrous. The predisposition to be born right-handed is lopsided. Greater than ninety percent of the world's population is born right-handed. Scriptures are loaded with references pertaining to the right hand, the idea of righteousness and it being synonymous with trueness, strength, rights, correctness, the right hand of God is the Son of God, on and on. Dexterity is the ability to perform tasks well with both hands coming from the Latin root dexter meaning of or on the right. Africa, the resonating tectonic template is a right-side profile of the human skull, exposed to the constants of the Sun, Moon, and other stellar influences, making our right-handedness a predominant feature. Are we right? Maybe?

People all have distinct fingerprints, like little sensors that look exactly like resonant field prints. They are the result of something beautiful like the touch of a lover, or the caress of the soft touch of sweet jazz. And yet, when little children or people shuffle their feet across a carpet and touch somebody, there is a little release of lightning, static electricity. The thunder comes when the cosmic parents smack their children. No!

Pounding in people's chests are their hearts, a reminder of a time before time when people did not have the means to evaluate reality without the original unity of being, which is still birthing the whole of creation at this very moment. The Breath of Life is the living word "I am." Breathe in. 'I.' Breathe out. 'Am.' Speaking the words "I am" in this way produces a resonating pattern. In repetition, the Breath of Life produces a vibrating field around the person, stimulating a relaxing feeling, much like a purring cat or a monk's intoning, "Om mani padme hom."

We have spoken of the chakras or the spiritual centers of the body where we sense an unseen reality. These chakras are circulating a glandular system of sensing reality through various resonating levels of interdimensional aspects. If you want to sense your chakras there are some known methods like whirling as a Dervish, laying in a hammock gently swaying in the breeze, or by meditating with your eyes closed.

Sound travels around and through the entire being body, into the ears. Their design is perfect, like the conch shell spiraling sound into a person's head though the tiny spiraling cochlea. When the tiny cilia, the hair follicles in our ears that make contact with the tectorial membrane vibrate, we hear the music of creation and the world.

Working construction, the crew and I would all start early in the morning and work late into the evening. Six, seven days a week. Ten, twelve hour shifts. During the winter the weather became brutally cold, freezing the lake out up the hill from me. One bitter morning I remember the wind blowing through all my layers of clothing. The warmth of my body could not keep the cold wind from blowing through my bones. I felt as much part of the inanimate reality as the concrete wall next to me. Bones are composed of calcium. The more they vibrate, the harder and thicker they get. Can human beings feel the sustaining life cycles in their bones? As diurnal creatures human beings wake up to the warmth of the sunlight and lose consciousness when the weather turns dark and cold.

Like a living stone, the skull is a hardened temple for protecting the sacred world within us, our brains. The skull houses the brain, which functions miraculously and processes information that a person's soul can learn, experience and never forget. What makes human beings unique is their skulls are unusually large for their stature and can store an infinite amount of ideas, memories, experiences, knowledge, and truth.

The theory of tectonic resonant evolution describes the laws that create human bones. The bones tell the truth. When people die, more bones are born, and to the ground they return to form the world for future generations. Do we exist in an iterative process to make more bones, skulls? Maybe? The only reason that humanity exists, in the end, appears to be bones and skulls. How morose. There must be more.

# CHAPTER 14: RESURRECTION

From Africa comes many of our religions, including, Judaism, Christianity and Islam. They all seem distant and out of touch a little bit with their own roots, even though they can count all of their forefathers back to Adam.

Let us consider the possibility that Jesus Christ's Crucifixion at Golgatha is not just a coincidence. Golgatha, according to the Bible, the New Testament, tells us in Matthew 27:33 "And when they came to a place called Golgatha, that is to say, Skull Place."

In Mark 15:22 "So they brought him to Golgatha, meaning "Skull Place."

And, in Luke 23:33 "And when they got to the place called Skull, there they impaled him and the evildoers one on his right and one on his left."

The Crucifixion of Jesus Christ at Golgatha directly ties into the idea of salvation, life, death and resurrection, where the centerpiece is the skull as an artifact of a greater image of life and death.

John the Baptist, while declaring the kingdom of heaven is coming and baptizing many people like Jesus Christ, is beheaded. In Matthew 14:10-11 Herod "sent and had John beheaded in prison. 11 And his head was brought on a platter and given to the maiden, and she brought it to her mother."

Tectonic resonant evolution is almost exactly that, a theory that we come from a head on a platter. Heads, skulls, and even mummies tie into the idea of resurrection. In fact, in Mark 6:16 and Luke 9:9 Herod fears that Jesus is actually John back from the dead. Jesus comes back to life from within a sarcophagus. Are we in the sarcophagus of God who will someday resurrect and the plate of Africa is about all that is recognizable of what was once the real head of the Most High God?

The Earth relies upon the life of Heaven for the sacred rain sustaining all life forms on Earth and in the waters. God's kingdom is reliant upon the life sustaining stillness of the Earth. Lightning and thunder are a direct energy exchange between that which is Above and that which is Below.

Thunder is considered the voice of Heaven that the stars interpret, echoing, resounding into the ground energizing it. Just as our voices echo in a cave, cavern, or canyon, thunder echoes in the sky above, resonating across the sky between the surface of the Earth and the limits of the sky. You can feel the Earth shake and vibrate when thunder booms. Sometimes during a storm, lightning goes from the ground up into the sky. Is Earth talking back and responding to the things the storm is saying. What language would this be?

Have you ever heard of balls of lightning? Sometimes the balls of lightning bounce off various layers of storm strata and land on the ground without exploding. Electricity, sizzling and crackling through the air, chases the ball as it moves about with an intelligence, sometimes starting fires, other times rolling towards or away from people. As suddenly as the ball of lightning appears it vanishes, sometimes in the accompaniment of thunder, other times with no bang at all. Some early woodcuts give an artists rendition of the balls of lightning as having actual faces, depicting awareness, as if the balls of lightning were alive with personality. Technically, balls of lightning fit the definition of unidentified flying objects, U.F.O.'s.

Parts of the description of the ark of the covenant implies that there is such an orb, communicating and advising the prophets. The description of the burning bush speaks of balls of fire, fiery red like glowing sapphires, later calling them sephirah. Orbs and crystal balls are part of the esoteric mystical experience, and even nowadays spiritual advisors claim orbs facilitate spiritual communication. Intelligent orbs of light have supposedly been photographed creating crop circles, designs of magnificent accuracy in fields throughout the world.

Is reality a living expression of the resonant field, an orb? Where did mankind acquire such profound ideas of God, a God that creates us in His likeness? The image is in the myths. Here is where God's face is visible. His soul, orb, or star is our sun lighting up the world.

The first myths speak of how death came into the world. When we die do our souls depart? Are they orbs? If Africa is actually a huge skull, or what remains of one then is the Sun the soul of that skull, departed, but not gone? Are balls of lightning like souls being caste back from Heaven to Earth?

Should the theory of T.R.E. have a subtext called the tectonic theory of resurrection, implying the Earth is a sarcophagus which serves as a resurrection chamber for God? God's soul, the Sun, resides within the orb of Earth's influence, disembodied outside His skull, Africa. In the center is where creation grows, expands, and recompresses back into itself, into a finer and finer crystallization of itself. Earth, the world, seems to be a containment system like a sarcophagus or a stasis chamber sustained by the planets, moon, stars, Earth, lightning, thunder and all other living things.

From the Earth, God takes eleven scoops out of North Africa and ten scoops from the Pacific Ocean which at the time is a crater of God's blood from His previous death. Taking the scoops and a single flea, God fashions a thermal-regulator, the Moon, to counter balance all of the activities that overheat the interior of the resurrection chamber, the orb, the world humanity lives in. When the conditions are right God's Soul, the Sun, the Dead, the stars and spirits reanimate into the template of the Orb. God is continuing to create creation at large, and doing whatever God does, and no prophet or seeker of knowledge ever knows it all.

For the divine feminine, the other part of the resurrection theory states we live within the orb of the Queen of Heaven. The soul of the Goddess is outside the skull of Africa. Her disembodied soul is the Moon and She is the Queen of Heaven, the lover of the King of Heaven, the Sun. When God nears Her face, head, Africa, to try to revive Her, this burns the land of North Africa, turning it into sand, like the Sahara.

Humanity only exists because of the space between the Earth and the Moon, which together is a matrix allowing for life to exist in all of its forms. When the Goddess resurrects, all of the land masses of Earth will form into Her living material body. All living creatures will resurrect in both the flesh and the spirit with vivid astral bodies, that are neither completely spirit nor completely flesh, but rather, immortal, once again.

Tectonic resonant evolution reveals the possibility of resurrection as a fundamental result of the world's resonance. Africa's skull shape is the progenitor of the human skull. Before God, comes the Mother Goddess. Nobody ever comes out of God's womb or nurses at God's bosom. The Goddess has the first crack at raising and teaching the Children of God. Earth gives birth to all of Heaven, the galaxies, stars and planets. Together the Father and Mother bear their child, creation, producing everything existing within the limits of the orb.

# CHAPTER 15: CONCLUSIONS

Today our common calendars are a result of the study of the stars, the planets, the Sun and the Moon, as they relate to the seasons, the designs of God and Goddess, holidays, and celebrations. However, I have not found any evidence that Africa was ever considered the center piece of any ancient or modern astrology. Recently the idea that aliens from outer space created us has gone more mainstream. The ancient astronaut theory looks to the stars for the source of their creation. Africa, the resonating continental template for humanity, the resonant field of creation and the world create humanity.

The design of the Tower of Babel trapped too much resonance from the Earth inside the tower, amplifying the resonance of the land under the tower, causing its collapse. What did the ancient builders learn? Puzzled by the collapse, ancient temple builders try a more squared and flattened ziggurat, temple style, with hanging gardens.

What is overlooked, is that the Giza Plateau was a mortuarium for a world soul journey through the Afterlife. There are three pyramids, three mortuaries, and three paths to the Nile. One path each for the Gods, Goddesses and their children, sending or sailing them all on their adventure on the cosmic ocean, the crossing of the celestial Nile, Milky Way of the starry world.

The Pyramids were to make sure the sarcophagus, ark, spirit, or soul had a safe journey, smooth sailing, through the Duat, all the way to the Great Resurrection Day, sounds Christian. Temple builders built the Duat encapsulating the idea, As Above so Below. The three Giza pyramids are the belt of Osiris, or the constellation of Orion. Below the pyramids resides the head of Osiris, as in the continent of Africa.

The first principle of the ancient Hermetic Formula for comprehending reality is: As Above so Below, as scribed in the Emerald Tablets. Genesis 1:7 clearly informs us of waters Above the firmament and waters Below the firmament. As Above, so Below. Did I really say that? Or is it a joke, so old that it's an old standby.

Resonating together, the waters above and below the firmament create the breathing mechanics for the world. It lifts up and drops down, oscillates, expands, and contracts back down. The surface of the land acts like a diaphragm and the sky's shell acts like lungs. A vortex, trachea, esophagus or throat opens and circulates the layer of gases throughout the living bio "sphere." The world's resonating field incarnates into an example like the tortoise. Heaven and Earth are living beings, and like all living beings exchange substance.

Clouds can only form within their resonant containment field, the troposphere, where all weather forms in the sky. The clouds descend from above into the lower portions of the troposphere, flattening out on the bottom. The Hermetic Formula teaches us "as above so below." If the clouds flatten out, like above, then below, the dirt beneath our feet, the Earth, must be like the clouds, flattening out, on the bottom, over years of sedimentation. The sky above the clouds is a dome and the surface of the Earth is flat. When it rains, it pours, like snow in a snow globe.

Studying mythology, anthropology and archeology for a long time, I noticed many of the myths come from the same story. As sacred knowledge is passed down from one generation to the next, sometimes the knowledge is lost in translation, such as the silent H dropping from many words throughout time. Survivors of repeated cataclysms build new monuments, sometimes over old monuments. There is no coincidence in the similarities of the clues left behind by our ancestors.

**Figure 11. The Word God**

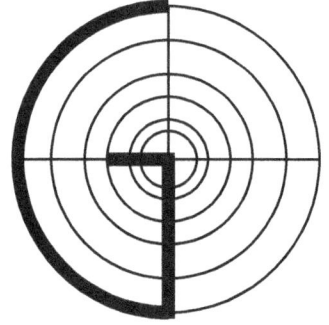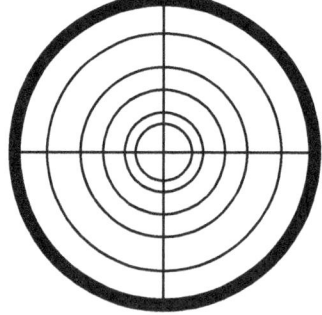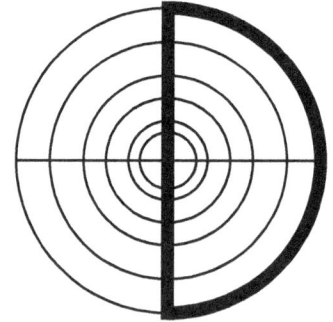

Creation's resonant field is displaced along the back of our friend the tortoise. On the belly side of tortoise resides the four corners of creation, the center, the beginning of creation, and the tortoise. The metaphoric mouth of Earth, Map of Atlantis, Hexagram 27 of the I Ching, Poseidon's Trident, and the Golden Lampstand, are all derived from the same teaching. Sacred geometry is the knowledge of the tortoise, revealing form, shape, the worlds dimensions and the word God.

Scientists, may see the profile of Africa as a source for the human skull image. The divine special creation theorists, believers in God, may expand their beliefs in their miraculous Creator to include the profile. I doubt the creationists and scientists will ever call Africa, God, Adam, Gaia or the Creator. I am not saying Africa is God, Adam, Gaia, Ymir, Tiamat, Osiris. What I am saying is that Africa is the resonating continental landmass responsible for shaping humanity's skull.

# AN AUTOBIOGRAPHICAL AFTERWORD

Well we said it was a theory. We were struggling to come up with a fancy equation to prove our theory but many of the variables depend upon two distinct and sometimes blending world views, Globe Earth and Flat Earth. The old model is geocentric with a dome stationary flat Earth as the center of the universe. Currently the model for Earth is heliocentric with a spinning globe. What we've done here is describe the variables that are independent of, and exist in, both world views.

All resonant fields can be drawn out into an image called a mandala. The first variable is O, a circle, the whole image, that which is created. Earth's image of natural frequency is given the variable E, and Creation C. The variable T best fits the resonant frequency of the tectonic landmasses, which already have an image, like Africa. Water, the variable W, has its own resonant field image and H is how you feel about it!

The result of combining the resonant field images of T, W, and E produces the image in the center of the whole of creation. For example the theory of tectonic resonance states that the human skull is generated by the tectonic resonating template of Africa. The equation would be the combination of the resonant field images of Africa, Earth and the Water level equaling the image of the human skull.

I have been studying the ancient art of making sacred geometric templates using the tools and the knowledge of sacred geometry for over forty years. Each template is a geometric interpretation of the knowledge compiled in the mass of numerical configurations of the common cosmic laws of the unity, duality, trinity, and so on.

My favorite study is the Mandala Effect. Mandalas are two-dimensional drawings representing different resonating fields. The Mandala Effect is when a mandala produces a three-dimensional effect a warping, a living breathing moving impression. Even though the mandalas are motionless, still, they seem to be alive.

Here is how I stumbled into the theory of tectonic resonant evolution. I was going blind from cataracts and the vision was fading so fast I had to finish up my most important templates, one showing how the universe may be and the other, the last I did before having the corrective surgery was a Fibonacci sequence swirl spiraling out from the center of a template of 56 circles joining around a common center equally spaced inside an outer circle. When I finished that last one I noticed that in the negative space left unmoving in the pattern I could put a skull and it might be really cool because it was nearing Halloween and the fluorescent construction paper I was using would be cool to look at under a black light and make for a good poster. So, I cut out a skull profile from a scrap of black construction poster board and put it in the empty space left at the center. It still did not look quite complete so I added the final touch of a pink fluorescent eye.

When I was finished I was listening to the television and Dr. Charles Stanley was preaching in front of the continent of Africa while walking back and forth in front of that image, when I decided to call my last art piece, "Africa." It was one of those moments where the gongs sounded. Then all of the bells went off and the lights flashed, the whistles blew and the horns, too. There may have been a bagpipe. I shouted Eureka and Hallelujah in hushed screams of private celebration because it was like finding an answer to reconcile the spiritual rift or tear in my soul created by the polarity of the fundamental themes of our origins namely: Am I a child of God, or just an animal, a higher ape? I suddenly found reconciliation, a middle ground, that was really ground, and suddenly I felt grounded spiritually and biologically. What is more is that when a hurricane spawns over the Atlantic ocean as a storm cell comes out of West Africa, the swirl of the Fibonacci sequence spiral pattern becomes a manifest phenomena in the hurricane's spiral pattern.

My quest was to find the truth, namely, where does the image of man come from? I know that everybody is on the same quest at the same time, to discover the truth, before the space between their ears is filled with manure. The manure is composed of all of the ideas offered as answers to the quest for self-knowledge that did not resolve the questions about the image and its origin. Our whole body of knowledge has no head. I think that maybe, we have found it.

After my eye surgery I spoke only to a few about it. Then I began re-researching the myths and legends and have found some valuable clues that certain ancient cultures know about resonance and apply their knowledge to the building of some of the ancient temples. If we look harder we can find more proof that ancient man was aware to some degree of what this theory is about, even though modern space age atomic warfare genocidal man knows nothing of this theory. God knows what mankind will do when they discover this theory. How will they see themselves? If this theory changes how they perceive themselves, how will they adapt to all of the insanity they are currently drowning in because this knowledge was obviously unknown to them?

The Gospel of Thomas, verse 56 claims Jesus said, "Whomever has come to understand the world has found (only) a corpse, and whoever has found a corpse is superior to the world."

One final thought: Resonant fields create us, and can also destroy us, there are two paths we can take. One leads to life. Tectonic resonant evolution proves that we are one and that we are all interconnected. What can be done to cherish this finite blessing given us by the Sustaining Infinite?